Pierre-Gilles de GENNES

A Life in Science

Pierre-Gilles de GENNES

A Life in Science

Laurence Plévert

Science Journalist, France

World Scientific

NEW JERSEY · LONDON · SINGAPORE · BEIJING · SHANGHAI · HONG KONG · TAIPEI · CHENNAI

Published by

World Scientific Publishing Co. Pte. Ltd.

5 Toh Tuck Link, Singapore 596224

USA office: 27 Warren Street, Suite 401-402, Hackensack, NJ 07601

UK office: 57 Shelton Street, Covent Garden, London WC2H 9HE

British Library Cataloguing-in-Publication Data
A catalogue record for this book is available from the British Library.

Cover photo: © Marian Schmidt/Rapho/Eyedea

Interior photos: private collection except p. VII middle and bottom © Philippe Nozières. P. X © Anne-Marie de Gennes. P. XI, XII and XIII top © Françoise Brochard-W art. P. XIII bottom © Henry Thurel. P. XIV bottom © Francis Apesteguy/Gam a/Eyedea. P. XV top © Thierry Boccon Gibod/ Sipa Press P. XVI bottom © Marc Fermier.

Despite our efforts, we have been unable to find the authors of some of the interior photographs. The authors of these photographs are requested to contact Éditions Belin.

ISBN-13 978-981-4355-25-4 (pbk)
ISBN-10 981-4355-25-9 (pbk)

Typeset by Stallion Press
Email: enquiries@stallionpress.com

Printed in Singapore by B & Jo Enterprise Pte Ltd

Contents

Foreword

Over the years, many an editor hit a brick wall in an attempt to persuade Pierre-Gilles de Gennes to write his memoirs: he was not remotely interested in producing an autobiography. He would rather "experience things than talk about them". Nonetheless, in 2005 he accepted my proposal to write his biography, "on condition that it talks about the science!" He formulated this requirement reflecting the meaning of his life, a life marked by his passion for physics.

He made the effort to interrupt his research — which he was pursuing as unflaggingly as ever — for long enough to talk about his memories in the course of some 20 interviews, which he carefully prepared. They were based on the notebooks he had been keeping since the 1960s, jotting down a word or two on his ideas, sometimes with a date and a colleague's name (he was not one of those scientists who kept meticulous "lab notes" on the progress of their work). He intended to use them to reconstruct the sequence of certain ideas, or perhaps to jog his memory, but he had always put off the moment, too absorbed in his current research. Do they contain any potential scientific gems? Whether they do or not, there is enough there to give the historian of science food for thought. Pierre-Gilles de Gennes also unearthed old diaries to retrace the thread of his appointments and travels, year by year. And finally, to make sure that "it talks about the science", he provided comments on his list of publications, pencilling a grey cross besides the ones "worth" mentioning — about one in ten — and highlighting mistakes more often than successes. He could accurately recall articles published 40 or 50 years earlier, in just a few cases wondering: "what exactly was I on about here?". He would sometimes punctuate his explanations, almost apologetically, with "it's a bit abstract". He also gave me access to his correspondence, kept at ESPCI (Paris Higher School of Industrial Physics and Chemistry), where he was the director for 26 years. When I worried that these interviews were taking up too much of his time, his secretary, Marie-Françoise, reassured me that he seemed to enjoy going back over all these memories.

But how can a book recreate the reality of a man as protean as Pierre-Gilles de Gennes? Everyone who met him was struck by his charisma, which is impossible to convey in words. He also radiated enormous kindness, as well as a distinction paradoxically allied with great simplicity, especially in his language: when stuck on some difficult problem, he wouldn't concentrate, but "worry away at it" or "chew it over". To shorten a demonstration, he would sum it up with a "hup-hup" and finish with "or something like that…" to take the edge off his words. But the most striking thing in his way of speaking was undoubtedly that "we", which, while it might be collective at the beginning of the story, could become individual from one sentence to the next, leading his interlocutor to wonder: *so who else is he talking about?*", before realising that in fact there was no one else. That was the intention: it wasn't a "royal we", but a "we" of modesty, making him just one of many.

Why dedicate a biography to Pierre-Gilles de Gennes? Other scientists certainly contributed as much as him to 20th century physics, but not all revitalised the subject with such panache; not all had his iconoclastic personality; and, perhaps unjustly, not all were awarded, as he was, the Nobel Prize, that incredible magic wand that brings a physicist into the public eye and which, within a few months, turned Pierre-Gilles de Gennes into a media icon. It was when he entered the courtyard of the Collège de France to find it swarming with journalists and TV trucks on the day of the announcement of the prize, that he realised that it would transform his life.

Amidst the avalanche of mail he received at the time, one particularly amused him: "*I'm not surprised that you have won the Nobel Prize, because I heard your mother talking to mine… before you were even born*", wrote a childhood girlfriend. A joke? Of course, but his mother was so convinced of her son's genius that she sometimes introduced herself with the words: "*Hello, I am the mother of the future Nobel Prize winner Pierre-Gilles de Gennes.*" He grew up on the pedestal she constructed for him by constantly telling him he was the best. It is true that he showed unusual gifts, but intellectual talent is not enough to explain a scientist's success. Pierre-Gilles de Gennes had other qualities that led him not only to revitalise physics, but also to introduce a modern way of reasoning by manipulating orders of magnitude and variables to simplify problems. Throughout his life, he set himself challenges, taking on new subjects — even at the age of 70 — every time he felt himself to be in a scientific rut, preferring demanding labour and the satisfactions of knowledge to the comforts of established success.

Neither historical nor sociological, this book has the modest ambition of revealing something of the great physicist and something of the human being (in particular, he was kind enough to lift part of the veil over his private life by passing on personal memories). For it is above all the man he was that made him "France's favourite physicist", a man whose freedom of thought is missed more with each day that passes. As far as possible, this book tells the story in Pierre-Gilles de Gennes' own words, but there is no way that it can convey the power of his presence. Although the text follows the interviews as closely as possible, it also owes its existence to those who accepted, always with great pleasure, to talk about the times spent in his company. I am extremely grateful to each of them for sharing some of their memories: their insights or testimonies have made this biography possible, like a photograph put together from scattered fragments. Some fragments are missing, some are perhaps in the wrong place, and the odd piece of sticky tape is visible, but at least it's there.

Pierre-Gilles de Gennes always stressed how much the American physicist Richard Feynman had inspired him: *"For me, brought up in the French way with a mindless respect for formalism, the game changer was the Feynman Lectures on Physics (...): for our generation, the message of that book, its critique, its outbursts, were a sort of Road to Damascus experience"*, he wrote in 1990.[1] Let us hope that he too will become a guide for future generations.

---

[1] Martine Poulain, *La bibliothèque imaginaire du Collège de France*, Le Monde éditions, 1990.

Acknowledgements

This book would not have been possible without the personal contributions of Anatole Abragam, Mireille Adam, Armand Ajdari, Claude Allègre, Vinay Ambegaokar, André Authier, Pierre Averbuch, Jacques Badoz, Sébastien Balibar, Jean-Pierre Beaufils, Henri Benoît, Jean-Claude Bizot, Julien Bok, Bernard Bourgeois, Thomas Boutreux, Édouard Brézin, Françoise Brochard-Wyart, Dominique Bomm, Christiane Caroli, Jacques Chemin, Jean-Pierre Chevènement, Daniel Cribier, Mohammed Daoud, Guy Deutscher, Anne-Marie de Gennes, Christian de Gennes, Estrella de Laredo, Élisabeth Dubois-Violette, Jacques Duran, Georges Durand, Jacques Friedel, Jean Gavoret, Sylvain Gilat, Robert Goutte, Marie-Alice Guedeau-Boudeville, Étienne Guyon, Françoise Hartmann-Boutron, Yvette Heffer, Hubert Hervet, Bernard Jacrot, Jean-François Joanny, Jacques Joffrin, Charles Kittel, Michel Laguës, Marie-Françoise Lancastre, Liliane Léger, Jean Léoni, Jacques Lewiner, Alexis Martinet, Jean Matricon, Barry Mazur, Lucien Monnerie, Marie-Caroline Müller, Philippe Nozières, Roland Omnès, Marie-Christine Picard de Gennes, Pawel Pieranski, Phil Pincus, Jacques Prost, Bernard Prugnat, David Quéré, Élie Raphaël, André Rassat, Francis Rondelez, Gilles Rubinstenn, Daniel Saint-James, Lisbeth Saraga, Georges Sarre, Bruno Schroeder, Jacques Signoret, Christiane Taupin, André-Marie Tremblay, Gérard Toulouse, Madeleine Veyssié, Jacques Villain and Claire Wyart.

Chapter 6 owes much to Pierre Billaud, who described to me the conditions in Reggane, in the Algerian Sahara, at the time of the first French nuclear explosions (I recommend that you read his account of the early days of the French nuclear deterrent on his website).

Jacques Friedel closely followed this project and I am deeply grateful to him for his intelligent comments. Françoise Brochard-Wyart, Guy Deutscher, Jacques Duran, Bernard Jacrot, Lucien Monnerie, Philippe Nozières, Francis Rondelez, Madeleine Veyssié and Jacques Villain gave me invaluable help in the correction of the scientific passages.

Denis Blaizot, Philippe Boulanger, Sébastien Carganico, Claire Guttinger, Pierre Lutz, Fabienne Longeville and Olivier Vaginay also helped me, as did Caroline Burnel, Yvette Firminhac, Viviane Fuchs and Catherine Kounelis from the ESPCI library (Paris Higher School of Industrial Chemistry and Physics), whom I would like to thank for their assistance.

I am also grateful to the scientific department at Éditions Belin, in particular to its manager and the editor of this book, for having supported the project from the start.

Finally, I would like to thank Armand Ajdari, whose 1992 classes on soft matter at the University of Rennes I still remember, and to whom this book owes much. I hope that he will not be offended if I repeat the words of his grandfather, Robert Buron, a minister under the 4th and 5th Republics, which provide such a good summary of what makes a good researcher: *"The taste of the fruit that one eats and of the wine that one drinks, the smell of mown grass and the scent of flowers, the sight of the ocean in sunlight or in cloud, the sound of vibrating strings or the crash of waves upon rocks, the almost disconcerting softness of fur, velvet or hide... but even more than those, the sense of excitement that comes from the logical explanation of a problem or phenomenon not previously understood, the discovery of a previously unknown field elucidated by a new insight (...), that is the inimitable and marvellous taste of life."*

Chapter 1

Childhood Wanderings

As he grew older, seeing his reflection in the mirror, Pierre-Gilles de Gennes would often be startled to find his mother's features in his own face. Day after day, the resemblance became more striking, to the point that those who had known Countess Yvonne de Gennes it became a matter of comment for. Through this disconcerting similarity, his mother remained constantly in his memory. A woman of conviction and strong personality, it is an understatement to say that she exercised considerable influence on his childhood, a time marked by his parents' separation and by war, yet one that he paradoxically remembered as happy — unless reticence prompted him to conceal its most painful episodes.

Madame Yvonne de Gennes came from a very wealthy background, the Morin-Pons, a great Protestant family with banking interests in Lyon. Originally from France's Drôme region, her ancestors had migrated to Geneva following the revocation of the Edict of Nantes, with *"just what they could carry on their backs"*. In 1805, a young man called Pons persuaded a number of Geneva bankers to help him set up a loans office in Lyon, which was experiencing an economic boom. It was the start of a banking dynasty: the Veuve Morin-Pons Bank — so called in tribute to the heiress who, in the late 19th century, managed to save the bank when a cashier had run off with all the money — would remain in existence until 1996, when it was taken over by Sanpaolo Bank. Pierre-Gilles' great-grandfather, Henri Morin-Pons, only worked there out of duty, preferring stamp collecting and the opera to loans and investments. In fact, he was the author of *Numismatique féodale du Dauphiné*, a reference work on stamps, and of romantic operas, the librettos of which remained in Pierre-Gilles' possession. Paul, the son of Henri Morin-Pons and Pierre-Gilles' grand-father, took things further: he never set foot in the bank, but lived off his annuities, subsequently occupying a position as Austrio-Hungarian consul in Lyon. His first wife was a very attractive Englishwoman, who bore him a daughter, Yvonne Morin-Pons, Pierre-Gilles' mother. Soon after little Yvonne's birth, however, her parents separated and took little further interest in her, her mother remarrying a good-looking cavalry officer whom she followed from garrison to garrison, and her father spending much time abroad. *"My grandfather explored the Balkans on horse-back in the 1900s. Then, he spent every summer with his new wife, a very young and ravishing Italian called Antoinette, nicknamed Nietta, on the banks of Lake Balaton in Hungary or in Venice. We still have old leather suitcases with the labels of the hotels they stayed in. They led a life that we would find hard to imag-ine today"*, smiled Pierre-Gilles de Gennes.

In his childhood and teenage years, Pierre-Gilles regularly spent time with this *"charmed"* family. *"My Aunt Madeleine had a superb house with extensive grounds, near Lyon. She was an ardent Bonapartist, and all the rooms were decorated with souvenirs of the Emperor. Since, as a child, I was very interested in the imperial campaigns, I got on well with her"*, he recalled. *"We also used to go to Switzerland, as guests in some splendid res-idences. I could draw up an unending list of cousins living by Lake Geneva. Actually, the house where the negotiations between the de Gaulle government and the Algerian resistance took place was a former family house of ours."* The boy was also sometimes invited to join hunting weekends in Sologne organised by his uncles. *"They drove old black Citroen jalopies at speeds that seemed vertiginous at the time. I used to love those hunts, run-ning along the streams with the dogs."* This family background gave him the deportment and discreet elegance of young men of good family, qualities he would never lose. As an adult, however, he distanced himself from this milieu, returning to it only on rare occasions.

The Great War

As a child, his mother spent the majority of her time in this great family, which took her in in the absence of her parents. *"But she felt abandoned and had an unhappy childhood"*, regretted Pierre-Gilles de Gennes. She married in May 1913 on an impulse, but the marriage did not last, which hurt her deeply. When war broke out, she immediately signed up to become a nurse. After six months of cursory training, she was sent to the front. *"She lived through the worst battles: Chemin des dames, Verdun… She was a woman of extraordinary strength"*, he recalled. She was tireless in her attention to the wounded soldiers, showing a devotion that commanded respect from the doctors. In 1917, in one of those campaign hospitals a few kilometres from the front, she met Robert de Gennes, a doctor who had been mobilised at the start of the war, immediately after graduating. He had great presence. He was a count, and his father, Paul de Gennes, a great Parisian physician, founder of the Boucicaut Hospital in Paris.[1] This doctor had married a Brazilian, Jenny Barboza Tinoco, who had borne him two sons, Robert, father of Pierre-Gilles, and Lucien, his uncle, both of whom became doctors. Robert had inherited his mother's tanned complexion and dark eyes. He had a soft voice and an attentive manner. The young nurse succumbed to his good looks and charm. They would meet during their breaks and make plans, but the war looked like it would last forever. *"They saw people dying around them every day"*, sighed Pierre-Gilles de Gennes. When the Armistice was finally signed in November 1918, they set up house in Paris, in a fine apartment on Avenue de Camoëns, in the 16th Arrondissement, close to Palais de Chaillot today. Robert de Gennes started a practice that quickly drew an Anglo-American clientele, since Yvonne de Gennes, with her British mother, spoke fluent English and served as receptionist. They married in December 1919, when she was 29. *"They had a happy 10 years together"*, according to their son. From 1930 onwards, however, the relationship began to crumble. Yvonne de Gennes, who

---

[1] See appendix on the history of the family.

could be impetuous and rigid, was difficult to live with. She was also very talkative and would drown those polite enough to listen to her in words. Patient as he was, her husband found her harder and harder to tolerate, and began to have affairs. She became unbearable. He could no longer put up with her incessant reproaches and considered ending the marriage.

Pierre-Gilles was born on October 24, 1932, but his birth did not mend the relationship. Realising that separation was inevitable, Yvonne de Gennes reacted in an unusual fashion. She abandoned her son, then a few months old, and set off on a long solo trip across Europe, as far as Turkey, to assuage the pain caused by the failure of her marriage. Later, she would justify herself by saying: *"I had no interest in looking after a child who was not yet able to speak"*, concealing the real cause of her departure. During her absence, the little boy was raised by a nurse in his father's house. On his mother's return, three years later, everything would change: she retrieved her son, never to let him go again, as protective as a mother wolf. From then on, Pierre-Gilles would live with her, first in Versailles, then in Rue Fantin-Latour in Paris's 16th arrondissement, a stone's throw from the banks of the Seine, and would only see his father for the odd weekend and short holiday.

Two Places, Two Worlds

Like many children of broken homes, the boy lived a life divided between two places, two worlds. Arriving at his father's flat on a Friday evening, in Avenue de Camoëns, he would be welcomed by the governess Maria, a corpulent, warmhearted woman, who was devoted to him. She would take his coat and let him run off to play with Blaise, Zoar, Gin and Fizz, the house's four lively fox terriers. Then, Pierre-Gilles would have supper and go to bed without seeing his father, who would be working late. But as soon as he woke on Saturday morning, the boy would dash into the living room for a hug. The doctor was an affectionate father and, if visiting patients, would sometimes let his son come along. The child would then climb proudly into the paternal car and, consumed with admiration, gaze adoringly at his father. *"His patients loved him. He was a serene and gentle man, who never lost his temper"*, recalled Pierre-Gilles de Gennes. *"He had a sort of Brazilian calm that I have perhaps inherited"*, he speculated. *"My secretary, Marie-Françoise, tells me that I taught her the merits of being laconic. I tend not to get carried away with words, except when talking about science."* When the visits were over, they would go home for lunch with Alice Conway, a beautiful American whom Robert de Gennes had married in 1936. Pierre-Gilles called her Aunt Alice and got on *"reasonably well"* with her. But she was a moody woman, and it was not a happy marriage. *"I know, because my father had notebooks where he recorded the scenes she caused, for example when she didn't get the table she wanted in a restaurant"*, regretted Pierre-Gilles de Gennes. On Sunday afternoons, they would walk in the Bois de Boulogne, where Pierre-Gilles would sail his toy boats in the lake: a grey fishing smack with a red sail, an elegant white yacht and a cruiser with a clockwork motor. When one day this cruiser got lost in the branches, Robert de Gennes immediately promised to buy him another one. *"As a child, I was showered with presents, spoiled by a father who didn't see enough of me. I can remember some luxurious Christmases"*, he admitted.

"With my mother, it was the opposite. It wasn't at all her style, which was rooted in a Protestant upbringing." Yvonne de Gennes was not a mother who found it easy to cuddle her son. There were neither kisses nor effusions of any kind. *"She had that slightly distant British style. I can also be like that sometimes, and people are occasionally surprised that I find it hard to shake hands. My mother kept a certain distance, but she wasn't cold. I didn't have the impression that my mother was unloving — I remember she used to warm me up by blowing warm air on my back — but she believed in certain values, for example it was important not to show emotion and to exercise self-control"*, he explained.

In fact, the atmosphere in the little flat on Rue Fantin-Latour was quite cosy. Yvonne de Gennes had not remarried, but they were not on their own, as she employed au pair girls to help her, for example Alla and Irène. *"In the 1930s, there were many White Russian émigrés in Paris. My mother took an interest in one woman, whose officer husband had died in Russia, and was living on her charms. My mother was concerned about her daughters, Alla and Irène, aged 17 and 18, and therefore helped them. They very often looked after me, for example on holiday in Royan"*, recalled Pierre-Gilles de Gennes.

The boy had no difficulty moving between his father and his mother. *"I don't remember suffering from their separation. It was my life from an early age so there were no sudden shocks. Nor did I have any reason to reproach either of them. My mother always spoke of my father with admiration. She told me a great deal about their shared memories, about the war or their expeditions — which we would find heroic today — in Andalusia or in Morocco, for example, where my father painted watercolours that we kept."* However, the war was to bring this life as a spoilt little Parisian to an end, depriving him of his home and separating him from his father, who would die in 1941.

Fleeing the War

On September 3, 1939, after Hitler's invasion of Poland, France declared war on Germany. The Phoney War began and lasted until the German offensive in France in the spring of 1940. *"I don't remember the day war was declared, but I remember an incident in the Phoney War. The English navy had given chase to a German battleship that was attacking British merchant ships in the Atlantic. It was called the* Graf von Spee. *It had taken refuge in Uruguay, before being attacked and finally scuppered by its own crew.*[2] *The adventures of this battleship, which I followed with my father, who had a keen interest in naval matters, left a strong mark on me as a little boy"*, he recalled.

At the end of 1939, Yvonne de Gennes was at her wits end. Her concern was not so much the German threat as her son's health. Now seven, he suffered from a cough, difficulty in breathing and pains in his chest. Robert de Gennes diagnosed pleurisy. *"My parents were very worried, because it was the same tuberculosis bacillus that attacks the pleura instead of the lungs. I heard them say: 'This child is going to catch TB, and there's nothing we can do to treat him.' I can still see my father, in*

---

[2] The Battle of the River Plate, on December 13, 1939, is considered to be the first big naval battle of World War II.

uniform, because he had been mobilised in a hospital, sticking a huge needle into me", shuddered Pierre-Gilles de Gennes. Following this spell in hospital, his parents decided to get him out of Paris, both for the benefits of a favourable climate and to protect him from the war. In early 1940, Pierre-Gilles and his mother moved into a hotel in Arcachon, a place famous for its warm air. In July, on a seaside walk, they heard the strains of military music. The Germans had entered the town and were parading. As they marched past, Yvonne de Gennes went pale. She squeezed her son's hand hard and pinched her lips. Pierre-Gilles was shocked by this reaction from someone who usually allowed none of her feelings to show. *"I don't know if any tears fell, but inside she was crying"*, he still recalled with emotion. War had come, and they were in the middle of it.

They returned to Paris, but stayed only a short time before leaving for the free zone. Three quarters of the population had left the city. More than 13 million refugees were on the move. Yvonne de Gennes and Pierre-Gilles took the train to Grenoble from the Gare de Lyon, accompanied to the station by Robert de Gennes. Pierre-Gilles hugged him before climbing aboard. It was the last time he saw his father. Robert de Gennes would die in 1941 of a heart attack, like his own father, Paul de Gennes, in 1918. *"My father's death affected me, but having not seen him for more than a year and a half, I probably suffered less than would a child who had lived by his side and suddenly experienced a big void. The void was already there in a way"*, he explained. He didn't attend the funeral at Montrouge Cemetery in Paris, as his mother did not want to take him across the demarcation line. Sixty years later, Pierre-Gilles de Gennes wondered how he would have got on with his father if he had lived. *"In fact, I didn't know him very well. Although he was always ready to give me a hug, he was an intimidating figure, a man with a 19th-century upbringing."*

Indulgent father he may have been, but he was nonetheless a man of his time with an attachment to tradition. *"He was a royalist, and I remember that he used to hang out a flag to mark the Festival of Jeanne d'Arc."* He insisted on his son acting fittingly. For example, one day when he caught him whistling in the house, he reprimanded him severely: *"Behave yourself. I don't want you whistling like a stable boy"*,

Pierre-Gilles reported. *"I don't know if we would have got on well, but I feel an enormous affection for him. His life was often sad, particularly in his final years. He was very lonely."* Indeed, as soon as war was announced, his wife Alice had *"wisely returned to America"*. He spent all his days at the hospital and sent his son postcards full of false cheer, telling how he had brought home a big stray dog, which had made the governess Maria shriek, because there was nothing to feed him. Robert constantly repeated that the separation was necessary, because he needed mountain air for his fragile lungs, and would finish his messages with a kiss for his "chubby chap".

Arrested

When the train reached Grenoble, Pierre-Gilles and his mother took a coach to Villard-de-Lans. Yvonne de Gennes had chosen to take up residence in this fashionable "health resort", frequented by the Austro-Hungarian Empress Zita and King Hassan II of Morocco. In the coach, Pierre-Gilles could not tear his gaze from the mountains. The Alps would become his adoptive home.

Pierre-Gilles spent some very lonely months in Villard-de-Lans. Even in Paris, he had had no friends or playmates. *"I can't remember any children being around. My parents were over 40 when I was born. I imagine their friends had older children"*, he speculated. At Villard-de-Lans, he had no companions either, and spent all his time in his mother's company. However, she had to go to Paris to settle "matters", such as the estate of her father, Paul Morin-Pons, who died in 1941, but also to make sure that there was no damage to her apartment, and was often absent. On these occasions, she left Pierre-Gilles in a children's home, of which there were dozens in the town, places for children with tuberculosis to convalesce. There was a family atmosphere in the home, but Pierre-Gilles kept to himself and hardly mixed with the other children. He put a good face on things during the day, but felt afraid on some nights in the dormitory. *"I would sometimes get into a panic. Then the monitors, very sweet girls, would come running to comfort me"*, he confessed. They would rock him in their arms and hum songs. He would write to his mother, asking her to come back, but she would stay as long as necessary, sometimes several weeks. Perhaps this explains certain traits he showed in later life. *"He needed to have people around him, he found it so hard to be alone"*, according to his wife, Anne-Marie de Gennes. Whatever the truth, he felt lonely in that children's home and learned to look after himself. This feeling reached its height the day he was summoned to see the director of the home, a doctor who loved to pamper his young charges. The boy was nervous and curious: the director never brought children into his office. *"I have some bad news for you. Your mother has been taken by the Germans."* Pierre-Gilles burst into tears: *"Is she going to be shot?"*, he asked. The director tried to reassure him, but the little boy was not convinced. Yvonne de Gennes regularly crossed the demarcation line without authorisation, but that night things had gone wrong. The guide had led her with others up an embankment, but on the other side they had run into a German patrol, which had arrested them all. The child lived in fear until his mother was released from prison a month later. Usually, their reunions were unemotional, but this time she hugged him for a long time.

"She had fought like a tiger, shouting at the Germans: 'I'm not afraid of you, I was in the Great War, I was at Verdun'", recounted Pierre-Gilles de Gennes, aware that the woman who raised him was a real mother courage figure. She stood fast against the Germans and then, from 1943, she would take risks to help the Jews. *"She was a dyed-in-the-wool Pétain supporter at the beginning of the war, but, like many people, radically changed her point of view when she became aware of what was being done to the Jews"*. She would carry sacks of food to Jewish families hidden in the mountains, passing the German guard posts without bothering to change her story from one week to the next: whatever the season, she claimed to be going to pick mushrooms. She was afraid of nothing, which was a cause for admiration, but sometimes of embarrassment too. *"She liked strolling around the old villages and, when she saw a fine looking house, she would say: 'Let's go and look, it must be very beautiful.' Being a child, I was terrified at the idea of going into people's houses, but she had no inhibitions"*, he recalled.

Not only did she lead by example, but she was as demanding on her son as on herself. She gave him an upbringing based on abnegation and self-denial, which could have a *"somewhat military side"*, he acknowledged. For example, she pushed him to finish what he started. *"She would get a bee in her bonnet. She had decided that we were to go on bicycle rides. I can still see myself crying away, trying to pedal my heavy bike up a hill, with the sun beating down"*, he remembered. It was out of the question to get off and push, so he reached the top in tears. In ways like this, she encouraged him to push himself to the limit in all circumstances, until the little boy developed a steely character and acquired a tolerance for hard work that would prove critical in his career, since it is often tenacity that distinguishes the exceptional scientist from the average.

An Unusual Education

Yvonne de Gennes also had very strict views on education. In Paris, and later in Villard-de-Lans, she did not send her son to school. *"I don't remember having heard her criticise the education system. But she was so proud of her son, more than mothers usually are... Perhaps she thought that he needed a special education?"*, he speculated. Indeed, she quickly became aware of her son's aptitudes, lively, curious, memorising everything with prodigious ease. At the age of three, he could express himself with astonishing maturity. *"I also know that I learned to read exceptionally early, because my mother constantly talked about it. But that's nothing to boast about. It's not a sign of profound ability. It's a bit like people who can perform huge calculations in their head"*, he commented. In any case, he spent his childhood constantly being told that he was the best and his goal was to come up to the mark.

So Yvonne de Gennes undertook to teach him herself. In Paris, before the war, she had nevertheless sent him to the Hattemer school — *"very posh"*, according to Pierre-Gilles de Gennes — founded in the late 19th century by Rose Hattemer, *"a tutor in several aristocratic and upper middle-class families"*. The school's former pupils included Prince Rainier of Monaco, Jean d'Ormesson and Jean-Paul Sartre, not to mention the physicist Jacques Friedel, who would play a big role in Pierre-Gilles' life. From the age of six onwards, Pierre-Gilles attended a few hours of lessons every week at 52, rue de Londres, in the 8th arrondissement, but he did most of his work at home, under the guidance of his mother.

"She was very focused on my education. When she was around, lessons took up 100% of the time. She read me books aloud, for example the history of the Imperial Generals." But the most important thing was for him to read himself, so that was his main activity. *"The other smart thing she did was to teach me English at the same time as French. Well, in fact, she didn't really teach me English, but she talked to me all the time in English and read me books in English, like* Three Men in a Boat *by Jerome K. Jerome, a very funny book that describes the adventures of three men rowing up the Thames. My grasp of English was a huge advantage to me in my work. When I went to Cambridge for the first time (as an engineer at CEA) in 1956, I felt completely at home. I found it just as easy to give lectures and talk to researchers as to understand their way of thinking."*

However, as he approached his 10th birthday, his mother felt that it was time to look for a school. But instead of starting in the normal French class for his age, she thought that he should go straight into Year 5, with children two or three years older. She was turned down several times, in particular by the director of the Gap college, for whom it was inconceivable that a child who had never set foot in school, however gifted, could succeed in a higher class, but she refused to give up. In Barcelonnette, she went to see the principal, a Mr. Barrans, who suggested: *"Would you agree to him taking a little entry exam? Let's say, a French test and another in maths. Then we'll know where we stand."* Although he answered the wrong question in one case (he described the mountains around Villard-de-Lans when the subject was: *"Describe your village square"*), the director accepted him into Year 5, impressed by his style. Pierre-Gilles was to experience his first taste of school, but his induction would not be without problems.

First School Term

In 1942, Barcelonnette was part of the free zone, under the control of the Italians. For this reason, the situation there was less tense than elsewhere. The Jews enjoyed relative freedom, although they had to submit to a rollcall every midday in the main square. The town took in many refugees, and accommodation was scarce. Yvonne de Gennes had to make do with a small two-room apartment in the main street of Barcelonnette, Rue Manuel. The accommodation was modest: the room overlooking the street did service as Pierre-Gilles' bedroom, whilst the room with a mountain view acted as living room and bedroom for Yvonne de Gennes (she curtained off a corner for her sleeping space). A wood stove was used for heating and cooking. A small basin with a zinc jug stood in a corner for washing, and Pierre-Gilles and his mother took turns to fetch water from the house's communal tap. There was a wood parquet floor, on which Pierre-Gilles would amuse himself by making chalk drawings of the Napoleonic battles described by his mother. Yvonne de Gennes, used to more comfort, would grumble about the conditions, and made a scene when she discovered that the walls of her son's room were cracked and had pipes running behind them. *"She convinced herself that gas fumes were coming into the room and demanded that the owners repair them. She got in a state, as she was always 'unreasonably' worried about me, for example about the risks of tuberculosis"*, tutted Pierre-Gilles de Gennes. Similarly, when she learned that a polio epidemic had broken out in Bristol where, as a teenager, he was spending part of his holidays, she immediately made him leave the city, lambasting the host family for failing to notify her sooner.

In Barcelonnette, Pierre-Gilles was impatient to start school, but the early days were tough. *"I was a loner, and I didn't have the same habits as the local boys. I spent the first few months fighting, but it was character building"*, he suggested. He then joined the Gravier gang, rivals of the Peyrat gang, named after the town's two districts. At the age of 10, for the first time, he had schoolmates, and even one friend, Gilbert Signoret.

About the school, he recalls times in winter when he had to arrive early to light the stove in the class, before lessons, in order to thaw the ink in the china inkwells. *"I can't remember if I was a good student, I probably was, but the main thing I remember is that I loved learning."* He hung on every word spoken by his French teacher, Louis

Blanchard, *"an archaeologist who before the war had worked on digs at Roman sites on the Black Sea"*, living in exile in Barcelonnette. The teacher recognised the boy's potential and became fond of him. He would play an important role in Pierre-Gilles' life.

Yvonne de Gennes was not wrong about her son's ability. Despite the three years difference in age, he got better grades than his class-mates, except in maths. *"My mother knew nothing about mathematics. It was the only gap in the education she had given me. So she wisely arranged coaching for me with an elderly woman in Barcelonnette, Miss Lèbre."* He quickly caught up and from then on his mother took no more inter-est in his grades, so insatiable was his hunger to learn. This was also the time when he began to draw, a skill in which he showed a certain talent. Yvonne de Gennes arranged drawing classes for him.

First Love

The rest of the time, he played like any other kid of his age. *"I spent hours playing marbles on Manuel Square, near an abandoned sand pile. I was pretty good"*, he boasted. *"From there, I would watch the Jewish rollcall, every day at noon."* He also enjoyed firing his catapult at the glass insu-lators on the telegraph poles, and more rarely at the legs of rivals in the Peyrat gang. *"I was very proud of that catapult — a rubber band was not easy to get hold of then."* In winter, all the children would ski together on Davin Field. They had to climb the slope on foot, skis over their shoulders, tie them to their boots with leather straps, ski down, then climb up again, and so on until the end of the after-noon. In summer, he loved going to the Chaup swimming pool, an old, more or less abandoned cement tank in the mountains. Every spring, it was filled by a small river that flowed nearby. Icy at the start of the season, the water would warm up as the weeks passed. Being stagnant, it would also be almost black by late summer. No one cared: it was a place where all the local youngsters went to play their favourite game of diving off a tree trunk that hung over the pool. Amongst the girls Pierre-Gilles met there was Martine Bour, who became a lifelong friend, and Dominique and Fanny, two nieces of Louis Blanchard, whom he would tease mercilessly by ducking them under the water. But there was only one girl who really took his fancy, Marie-Thérèse Paret, *"very pretty, with her cute pigtails"*, he recalled. Alas, his sweetheart's parents were not pleased to see this skinny lad, with his shapeless shorts and the catapult in his belt, hanging around their daughter. Deprived of her company, he had to make do with sending her love notes.

Yvonne de Gennes did not always give permission for him to spend his afternoons at the pool, but he would ignore the ban and swim in his underwear. *"Sometimes, it wouldn't be dry by the time I got home, so I would put my shorts on over the top and cycle home. However, a big wet patch would make my disobedience all too obvious. I can still see my mother chasing me around the bed with an umbrella, and me trying to dodge the blows."* The boy often found himself at odds with his mother. Although she exercised a steely authority, he was perfectly prepared to oppose her. *"Even back then, I didn't necessarily do as I was told."* For example, Pierre-Gilles would ignore the lights-out at nine rule and secretly read in bed, covering the bedside lamp with a sweater. One evening, the sweater caught fire. Nothing serious, but it earned him a severe ticking-off from his mother. He waited stoically for the storm to pass. It was an approach that he would adopt all his life in dealing with conflict.

One apparently minor incident began the little boy's emancipation from his mother. On a walk one day, she and Pierre-Gilles came to a farm where a dog ran towards them, barking threateningly. Recalling his mother's advice never to show fear when threatened by a dog, but to look it straight in the eyes, the lad continued to advance, staring it directly in the eye. However, after he had taken three steps, the beast sprang forward and bit him. It was a painful lesson that his mother's rules could not necessarily be relied upon. From then on, he would seek to make up his own mind about things. What his mother lost in authority, he gained in critical thinking. *"I didn't become independent simply to oppose my mother, but also because her upbringing itself encouraged me"*, stressed Pierre-Gilles de Gennes. *"I remember another episode that left a mark on me. I was around five or six. I had climbed a wall, not very high. When the time came to get down, I was afraid to jump. An old gentleman came up, a retired general type, and started to lecture me about how little boys should never be afraid. My mother didn't say anything, but I could sense that she agreed"*. He also had to look after himself when she left him once again to go to Paris, making arrangements for him to go to the teacher's family only at mealtimes.

When the Germans occupied the zone in September 1943, the atmosphere changed and the arrests began in Barcelonnette. The Jews took refuge in the mountains (this was the time when Yvonne de Gennes began to bring food to several families every week). In the spring of 1944, the resistance launched a major offensive and took control of the sector. There was jubilation in the town, with the inhabitants picking bouquets of flowers and decorating the streets. Pierre-Gilles joined a group at the edge of town, watching the partisans trying to dislodge the Germans from a manor house. *"The partisans surrounded them and the manor house was burned"*, he recalled. It was a scene that left a vivid impression on him. In the days that followed, large numbers of troops were parachuted in. *"We kids were very keen to help fold the parachutes"*, he remembered. Unfortunately, the Ubaye Valley where Barcelonnette was located was a strategic route between France and Italy and, three weeks later, the Germans sent a tank battalion to retake the town.

The reprisals were terrible. The Germans occupied the school. The resistance fighter Émile Donnadieu was arrested, tortured and then shot in the school yard. Louis Blanchard, the French teacher who was close to Pierre-Gilles, also a member of the resistance, had to go into hiding with his family. Every day, the Germans shot five people at random. *"One morning, I had a real fright. My mother wasn't there and I was going for lunch with the primary teacher's family. As I approached the school, two Germans saw me coming, pointed their machine guns at me and pushed me against a wall. I thought: "This is it: I'm going to be one of the five."* The soldiers held him against the wall like that, legs shaking, for long minutes. *"In fact, they were going to execute the five victims on the other side of the wall. They didn't want a little boy to witness the scene"*. He heard the shots and the soldiers let him go. Pierre-Gilles fled without further ado.

During these weeks of reprisals, the boy lived in fear, particularly the day when his mother, learning that the Germans had decided to burn the resistance HQ, which was in fact the house of Gilbert Signoret's mother, went to the Kommandantur and enjoined them with raw courage: *"Don't burn the house, that poor woman and her child have nothing to do with it."* Back at home, Pierre-Gilles was convinced that she would be arrested and be selected as one of the five to be shot the next day. But in the end she came back safe and sound and the house was spared.

American Airmen

In August 1944, after the landing in Provence, the Americans arrived in Barcelonnette. Liberation had come. *"I was over the moon because, every morning, I would cycle over to see the Americans, who had set up camp around Barcelonnette . They were still shelling the forts occupied by the Germans. When they fired, I got out of the way."* He made friends and exchanged addresses with American soldiers from Brooklyn and New York, promising to write. He would lurk around the Piper Cubs, the U.S. Army's little spotter planes. Noticing his interest, one day the soldiers invited him up for a flight. Unable to believe his luck, the boy had his first taste of the air. *"Just a short spin, but I was very proud"*, he recalled.

The relationship with the American soldiers was tarnished by one unfortunate episode. Pierre-Gilles had got involved in trading. *"I used to find eggs in the surrounding farms and swap them with the soldiers — they were pleased to have fresh eggs — for rations or part rations, which contained chocolate and biscuits, things we dreamed about. We weren't starving, but we didn't have a lot to eat. In general, it used to work pretty well. But there was one occasion when it didn't..."* That day, no one wanted to do business with him and he found himself stuck with his eggs. The temptation was too strong, and he tried to steal a ration pack, but got caught. He was deeply ashamed, but a kindly soldier stepped in and agreed to make a trade, and he was let off.

Barcelonnette was free and it was a time of rejoicing, but also for a settling of accounts. The FTP (resistance fighters) would hunt down collaborators, beat the men up in the street and shave the heads of the women and put them on display. Pierre-Gilles was shocked that the heroes of the Resistance should behave so cruelly. He talked about it to Louis Blanchard who, without condoning them, tried to explain their actions, but the boy was not prepared to listen. This sense of rebellion coloured his emerging political awareness, and was reinforced by another episode in 1946. *"My parents, who were so involved in the 14-18 war, were deeply shocked by the 1939 defeat. As a result, I thought that Paul Reynaud was useless, and was convinced that he would never be seen again after his contribution to the disaster, his failure to organise a defence. I was naive. After the war, he re-emerged, managing to get elected as an MP. As a result, I became very mistrustful of politicians and speeches. I had the impression that they were a house of cards and I was not entirely wrong."* All his life, he would be wary of politicians in general and of the communists in particular, whose ideas, however praiseworthy, could provide a cover for dubious practices, as the FTP showed. At the time of the mass post-war student political movements, particularly in favour of the Communist Party, he would deliberately remain at arm's length from the militants.

Return to Paris

Yvonne de Gennes closely followed every step of the progress of the Liberation forces. In September 1944, as soon as she learnt that the route from Barcelonnette to Paris was open, she decided to return. She packed their bags quickly so that Pierre-Gilles would be back at his desk at the beginning of the school year in October. She found a car, which dropped them off on the outskirts of Lyon. The city had just been liberated, and Yvonne de Gennes and her son crossed it on foot. *"There were still corpses in the streets. Many of the bridges had been destroyed and we couldn't get through by car. We used a railway bridge that had been dynamited, but not entirely destroyed. It was tilted at a 45 degree angle. I can still see us with our suitcases, waiting to cross..."* There were many refugees heading home like them, and they had to take turns using the bridge. Certain sections were dangerous. Pierre-Gilles had to fend for himself, while his mother walked in front of him without batting an eyelid. The rest of their journey passed without incident.

Yvonne de Gennes was happy to return to the comfort of her apartment and her social life in Paris. Pierre-Gilles went into Class 3 at lycée Claude Bernard in October 1944, but the classes were transferred to lycée Janson de Sailly during renovation of the buildings at lycée Claude Bernard, which had been occupied by the Germans during the war. Once a week, he had lunch with his aunt Edmée Colomb de Donnant, who lived in a cosy apartment on Avenue Victor Hugo, not far from the lycée. He would chat easily in his aunt's lounge, and charm her guests with his wit and general knowledge. In this time, he re-established contact with the family he had not seen for five years. For example, he saw Lucien de Gennes, an eminent professor of medicine, who lived in grand style with his wife, Renata de Gennes, reputedly one of the most beautiful women in Paris. He went back to see his family in Switzerland and was delighted by the nylon stockings worn by his young cousins. A few weeks before, he had been splashing around in the streams of the Ubaye Valley, and here he was now, mixing casually with the cream of society.

In Paris, nevertheless, day-to-day life was still hard. *"It was still a wartime atmosphere. At night, we heard shots, guards firing at people who had come to steal coal from barges travelling up the Seine"*, he recalled. However, the insecurity was nothing like what he had experienced in Barcelonnette. Rationing was still in place, and meals were frugal. Already tall, Pierre-Gilles was almost thin. Little by little, though, life began to improve. On 8 May 1945, the German forces surrendered. At this announcement, Pierre-Gilles persuaded his mother to go to the Champs-Élysées. The street was awash with people. *"I remember an immense crowd and a sense of extraordinary joy"*, smiled Pierre-Gilles de Gennes. The war was truly over.

During his year in Class 3, the boy got very good results, good enough to go into Class 2. However, his mother thought that at 13 he was too young for the lycée. She took advice and reached a decision: Pierre-Gilles would take a "sabbatical year" and would only move on to Class 2 at the beginning of the next school year. *"It was a very bold move on her part, not at all the way most parents think"*, he commented. In the meantime, she took her son and Gilbert Signoret, his friend from Barcelonnette, on holiday to the very chic spa resort of Perros-Guirec, in Brittany. At the seaside, the two boys could at last experience the carefree fun that they had missed as children of the war.

Chapter 2

Birth of a Vocation

Up to this point, Pierre-Gilles de Gennes had absorbed only his mother's literary culture. Now, his interest began to turn more towards the sciences. He discovered physics and chemistry during his "sabbatical year", browsing around the Palais de la découverte [science museum], and from 1947 onwards his true vocation would begin to emerge at the lycée.

His first year as a teenager was a strange one! In October 1945, at a time when his schoolmates were going back to school, he stayed home. How could he feel other than "different" with this divergence from the beaten path? He had no imposed timetable and could spend his days as he wished. Nonetheless, Yvonne de Gennes always made sure that he had something to do. Indeed, she could not bear him wasting time. "*My mother was always busy, always doing a hundred and one things*", recalled Pierre-Gilles de Gennes, who inherited the same trait. "*He was incapable of doing nothing. For example, he would never go for a drink and a chat on a cafe terrace, unless he had no choice. He did everything thoroughly and was always doing something*", recounts his close friend, the American physicist Phil Pincus.

Yvonne de Gennes expected her son to use the year to improve his general education. For her, nothing was more important than education. So she constantly encouraged him to read, while monitoring his reading matter. For example, she refused to let him read the adventures of Sherlock Holmes, even in English — "*No point wasting your time with pointless novels*" — and would instead suggest novels like *Maurin des Maures*, by the French academician Jean Aicard, or biographies such as *The Life of Pasteur, The Marshall of Ségur* or the *Memoirs of General Baron de Marbot* (a trifling 650 pages). She also recommended the works of the Duchess of Abrantès, although she considered certain pages to be unsuitable. She would carefully pin them together to stop him reading them. But as soon as she turned her back, obviously he would go straight to these pages with their suggestion of forbidden fruit, only to be disappointed, since the passages were never as racy as he hoped. All in all, at the age of 13, he had the reading habits of an adult.

His mother also arranged for him to have a few lessons with Louis Blanchard, recognising his merits and his bond with Pierre-Gilles, although she disapproved of the historian's "*very leftist*" ideas, diametrically opposed to her own. "*She had a very conservative and WASP side*", acknowledged Pierre-Gilles de Gennes. The boy saw Louis Blanchard every week. "*We would often dive into the Louvre, where he would give me a sort of lesson in art history.*" They would spend hours wandering around the museum, going from masterpiece to masterpiece, Pierre-Gilles hanging on his tireless teacher's every word. Then they would go back to Louis Blanchard's home, near Rue de Vaugirard, and continue their discussions over dinner. Pierre-Gilles became a fan of the Louvre. He would be a regular visitor throughout his life, sketchpad in hand, even organising meetings and appointments by the sculpture of the *Winged Victory of Samothrace*.

During this strange year, Pierre-Gilles haunted the museums of Paris, experiencing his first encounters with science at the Palais de la découverte. "*The Palais was very dusty after the war, but I just didn't notice*". Indeed, he was captivated by what he discovered. "*What I found most spectacular was the experiments in optics, in which mysterious figures emerged out of the gloom*", he remembered fondly. He would then immerse himself in books to understand how the light waves combine or cancel each other out to form these figures, his first solo introduction to the theory of interferences. The electricity experiments also fascinated him. He never tired of seeing the spark flashing between two electrodes in a glass tube. He explored the museum from top to bottom and would spend hours in the mathematics room, unearthing entertaining problems to solve the way other children might enjoy a treasure hunt. The moment a mathematical riddle or physical phenomenon aroused his curiosity, it became a challenge to his intelligence and he would not rest until he had got to the bottom of it. He would browse through the books and celebrate when he found the solution.

His year of freedom was coming to an end and, with it, the final remnants of childhood. "*Coming out of the Palais de la découverte, I would still stop at the Tuileries Lake to sail the boat my father had given me before the war. I became very popular, because there were a lot of kids younger than me whose boats would get stuck in the fountain in the middle. Because I was tall and wore shorts, I could go and fish them out of the lake.*" Indeed, in that sabbatical year, which his mother saw as a "*year of maturation*", he grew both in stature and in character, attaining a rare degree of independence for a teenager. Yvonne de Gennes never needed to nag him to work, so keen was he to study for himself. That year also consolidated his self-confidence, which his mother's unconditional admiration and his teachers' earlier praise had already helped to develop. From this time on, he would always display "*the self-assurance of people who are conscious of their worth and the modesty of those who don't need to assert it*", observes his friend Madeleine Veyssié, a lab director at Collège de France.

A Studious Teenager

He returned to school in October 1946, going into Class 2B (Greek and Latin options) at lycée Claude Bernard, an imposing building near Porte d'Auteuil. It was not the grandest lycée in the 16th arrondissement, but almost, its intake mostly boys of good family. Plus-fours and sports jackets were the norm. Pierre-Gilles was one of the awkward ones who would not wear a tie. When the bell rang, the students would form ranks before going into class. Pierre-Gilles topped his schoolmates by at least 4 inches, despite being two years younger. He would soon top them in his grades.

From the start, Pierre-Gilles became friends with four boys. Boris Bespaloff was a thin, lanky lad of Russian origin, who resembled a large yellow grasshopper. He could be very funny with his cynical and world-weary manner. Bruno Schroeder, grandson of the writer André Lichtenberger, author of *Mon petit Trott*, a children's classic of the time, was the literary one of the group. He was a Protestant and would attend confirmation classes with Pierre-Gilles. Jacques Chemin, the least studious of the four, liked to play the black sheep. Finally, the serious and hard-working Bernard Prugnat was the closest to Pierre-Gilles in his year. He was as short as Pierre-Gilles was tall, but his match in intellect. For this reason, Pierre-Gilles agreed to work with him, whereas he would — gently — turn down any other student who wanted to revise with him or asked him for tips. Pierre-Gilles and Bernard would spend afternoons revising together. Pierre-Gilles kept thorough notes on record cards, highlighting the important points in colour. He was already highly organised, as he would be throughout his career: many a researcher would be surprised by his meticulous notetaking, his impeccably tidy — almost empty — desk, his dozens of neat document files, where he would be able to find the article he wanted in less than two minutes. During revision, the friends would amuse themselves inventing mnemonics. They also loved setting each other riddles, like this one, sent by Pierre-Gilles to his classmate: "*[Here is] a very amusing probability problem that I did the other day at my uncle's instigation. Take a needle of length one, and drop it multiple times on a flat surface covered in equidistant parallel lines (d = 2). What will be the ratio between the number of times the needle falls across one of the parallels and the number of times it doesn't? Demonstrate, assuming that the needle has zero thickness, that this ratio is 2 to x. I was lucky enough to spot the trick immediately.*" Their other big game was to improvise tirades in Latin, as in this letter that Pierre-Gilles wrote from the mountains, where his mother, always concerned about the fragility of his lungs, would send him for regular stays.

"My dear Bernard,

Thank you for the clever, wordy, satirical, ironical, documented and erudite epistle in which the day before yesterday I had the joy of reading the multiple events, reasonings, developments, hypotheses, metatheses, syntheses, syncopes, apocopes, transitions, conclusions and exclamation marks. With your distinguished Latinist's eye and ear you will have no doubt appreciated the harmony, the cadence and the progressive rhythm of the admirable ternary group formed by the final words of the previous sentence. Firstly, "transition", the first term, attacking note, appearance of the "ion" consonance; secondly, "conclusion", a term whose meaning demands that it should be placed second, the weak beat of the musical period; thirdly, "exclamation mark", emphatic fall, final fermata, *whose imposing allure contrasts with the lightness of the idea expressed. I draw a final line under this blablabla, in which I show with such natural ostentation the advantageous opinion I have on the subject of my ternary group technique, that group which my imitator Cicero and I have raised as conquerors on the horripilous fortress of the oratorical genus. Your description of Claude Bernard's revolutionary days*[1] *aroused a lively interest in me.* (...) Mirore commodo sensor claude bernadistas lessavit greva commensata cantare vociferaque vocitus (...) chansonnetam macabe.

[Translation: I appreciate how Claude Bernard's censor, the strike having begun, allowed the famous maccabee's song to be sung and roared, in voices as tuneless as they were loud.]

From school to the Alps; I've been skiing here for three days. I stayed in Paris until then, since the doctors didn't want to let me go, plus the strikes. (...) Anyway, I must hurry up and finish, with a very affectionate handshake."

Competition between the two friends was intense — which would get the best grades? They alternated at the top of the class. Whilst they were well matched in grades, Pierre-Gilles stood out for his wit and his capacity for self-expression. One day, in a Latin lesson, the students discovered the historian Flavius Josephus and were laughing

---

[1] In December 1947, the lycée teachers went on strike and the students were sent to the prep room, where they chanted the song of a maccabee. It was the event of the year.

about his name. *"It's as if he called himself La Rochefoucauld — Roukhomovski"*, observed Pierre-Gilles, Roukhomovski being one of the pupils in the class. The latter glanced up, surprised to hear his name, and the class burst out laughing at his nonplussed look. Pierre-Gilles did particularly well in French. *"The teacher would frequently read out extracts of his essays, whereas though I was first in composition, he never read out my work. Pierre-Gilles had a personal touch which made him stand out"*, recounts Bernard Prugnat.

Pierre-Gilles became fond of his French teacher and would go to visit him, a fact he kept a closely guarded secret from his schoolmates. *"The same thing happened with Jacques Nathan as with Louis Blanchard, we became friends and we remained in contact"*, he explained. There was no suggestion, however, that he saw them as substitute fathers and he expected neither affection nor support from them, seeking out their company simply for their erudite conversation. *"For example, I was never tempted to ask them for advice"*, he asserted. Annie de Gennes confirms that the relationships were more intellectual than emotional.

In his final year, he showed the same facility in philosophy as he had previously in French. In classroom assignments, Pierre-Gilles would read the question, think for five minutes, then write one page, or two at most, and put down his pen. He would discreetly take a book out of his briefcase and read until the end of the test. The other students would have trouble finishing the essay in the two hours allocated. And Pierre-Gilles would get the best grades. At the age of 16, he was capable of producing a lively synthesis and analysis of the complex thought of Kant or Heidegger.

Humanities or Science?

Despite his aptitude for the humanities, he shifted towards the sciences at the age of 15, after failing to win a prize in France's national competition. The day after the Latin translation test, Jacques Nathan asked him: *"How did you translate this sentence?"* The teacher was horrified when he heard the answer: *"You're done for, you fell into the trap!"* Pierre-Gilles did not care. By contrast, only getting an average grade for his essay on "The Sublime" stuck in his throat. He said nothing about it, but was disappointed. He realised to what extent the assessment of an essay is subjective and how arbitrary success in the humanities can be. *"Until then, I was as interested in French and Latin as in science. And I loved writing"*. Apart from his schoolwork, in fact, he wrote short stories, slightly gloomy in content, in notebooks, which he would then type out, but *"they're not worth mentioning today"*, he laughed. *"Since I had read a lot and liked writing, becoming a writer seemed natural. I gave up the idea, because I found that a fog of artistic imprecision reigns over such activities, and people can get away with anything. I stopped thinking of the humanities as a career."* His vocation was emerging. *"The sciences attracted me both for their mysterious side (the fact that there are things we don't understand) and for the robustness of the result. I undoubtably acquired my vocation from my teachers at lycée"*, he concluded.

His maths teacher, Camille Lebossé, was the author of the Lebossé-Hémery textbooks. Despite appearances, he was a fantastic teacher. *"He had the gift not only of making his students understand algebra and geometry, but also of making them enjoyable"*, recalls Bernard Prugnat. Between demonstrations, C. Lebossé would sometimes talk about his war experiences, in particular his five years in a prisoner of war camp. *"Don't believe that the Germans were unaware of the existence of those camps. When we were behind the barbed wire, thin and starving, whole families would come and stare at us from outside"*, he told them. Pierre-Gilles admired him all the more for having overcome these experiences. Later on, he would always draw attention to the merits of researchers from countries such as Russia or Israel, who had achieved a level of excellence despite hostile conditions, stressing how aware he was of the favourable circumstances of his own beginnings. The other teacher who gave him a taste for the sciences was Monsieur Roger, an *"excellent"* teacher of physics. Although his lessons were in classical physics — *"I don't think I had even heard of Einstein"* — Pierre-Gilles revelled in the problems, which he found *"fun"*: "A rope without mass is hanging from a pulley. Two monkeys of equal mass are hanging from each end, at the same height. One stays where it is, the other starts to climb at a constant speed. What happens?"; "Why does a mirror reverse left and right, but not top and bottom?"; "When an ice cube, floating on the surface of a completely full glass of water, melts, does the water overflow?"... He excelled in solving them. The decision was taken: he would study science.

First Mentor

Becoming aware of his infatuation with the sciences, his mother sought advice from Edmond Bauer, Professor of physics at the Collège de France and a member of the Academy of Sciences, a former student of Jean Perrin and Paul Langevin. "*I don't know how she knew him. She had an initial meeting with him, and asked his advice. I had little idea that the man before me had been the leading light of physical chemistry in France since the 1930s, and one of the few Frenchmen to have understood quantum physics from the start*", explained Pierre-Gilles de Gennes. Bauer became a family friend and mentor to Pierre-Gilles, and influenced him greatly.

So when Edmond Bauer offered him a copy of his 1949 book *L'Électromagnétisme hier et aujourd'hui* [Electromagnetism Yesterday and Today], Pierre-Gilles took it into his head to redo the demonstrations. And when he came across the formula that links the angle of a needle hanging from a wire to the twisting torque applied, he started on the underlying calculations. "*I saw it as a challenge. I told myself that I had to succeed in doing the calculation, which was not easy, because it involved triple integral equations, which I hadn't yet studied.*" He managed to reconstitute the formula and the pleasure it gave him was as intense as if he had climbed Everest.

Edmond Bauer was also behind another important encounter between Pierre-Gilles and science. When he learned that the 15-year-old would be spending the summer of 1948 in Bristol, he suggested that he should visit his Italian friend, Giuseppe Occhialini, a specialist in particle physics.[2] Pierre-Gilles wrote to him, and a few days later

---

[2] Giuseppe Occhialini discovered the n-meson or negative pion in 1947, with Cecil Frank Powell (Nobel Prize for physics in 1950) at the H. H. Wills Laboratory.

was admitted to the summit of the Victorian tower where the illustrious physicist worked. "*It was my first encounter with real physics. All the windows were covered with black curtains. It was completely dark and there were ten-metre long photos of cosmic rays. You could see new particles appearing in them. Child though I was, the physicist very kindly explained everything. I didn't understand a word, but the surroundings, those big, dark, empty rooms and those immense photos left their mark on me. I saw him again in the 1990s, but he didn't remember me*", regretted Pierre-Gilles de Gennes.

The importance of the meeting with Edmond Bauer went beyond these "firsts", in the sense that he even influenced Pierre-Gilles de Gennes' research. "*I think he suggested research that would be interesting to do, and above all he gave him a certain idea of research, which left its mark. Edmond Bauer had fully assimilated quantum physics and was applying it to macroscopic problems, which was also the approach Pierre-Gilles de Gennes took. In addition, a lot of the research done by Pierre-Gilles de Gennes matched the sorts of topics that interested Edmond Bauer and are in the spirit of physical chemistry*", judged Jacques Friedel, a French physicist specialising in the theory of metals, who would be one of Pierre-Gilles de Gennes' teachers.

Jazz in Saint-Germain-des-Prés

Although he was a bit of a "*geeky*" teenager — "*I read and worked a lot. I also listened to the radio, which used to broadcast good plays and short stories*", he recalled — he nevertheless had teenage leisure activities. For example, ice skating. He learned to skate with Nietta, his maternal grandfather's second wife, and was a regular at the Molitor skating rink. "*There was a French champion who used to train there, whom I admired a lot: Jacqueline DuBief. I found her very beautiful and I haven't forgotten her*", he smiled. "*I would also often rollerskate across Paris. At the time, it wasn't dangerous, because the traffic was light. I would even to get on the bus in rollerskates. The conductors weren't that keen...*"

Catching the jazz bug, he took up the guitar in 1948. "*He learned to play very well without lessons. It was amazing how he succeeded in everything he undertook*", comments Jacques Chemin. In 1949 they formed a New Orleans style jazz band with Pierre-Gilles on the guitar, Jacques Chemin on drums and Dominique Barbier on the clarinet, and rehearsed for two years "*more at my place than his, as his mother wasn't very receptive to that kind of music...*", adds Jacques Chemin. The latter's room became the hub of local youth culture. "*I had a 78 RPM gramophone and my friends could come and go through the window*", he explains. "*I used to spend my whole time in that place. We would get together to listen to jazz. We would set each other riddles, argue, criticise, and then play a bit of music. Great times*", recalled Pierre-Gilles de Gennes.

Jacques Chemin knew all the best places and, on Saturday nights, he would take Pierre-Gilles to jazz clubs like Club Saint-Germain or to surprise parties. They would drink strong white wine and smoke yellow gitanes. "*It wasn't anything clandestine in those days. It wasn't an expression of rebellion either. It was simply what we did*", noted Pierre-Gilles de Gennes. They would listen to music and dance, except Pierre-Gilles, who was not a dancer. He was also careful to avoid pointless discussions, as he could not stand people who liked the sound of their own voice. "*There was no one like him for finding a put-down for anyone unwise enough to utter a cliché, and he could be very cutting, even with his friends*", recounts Jacques Chemin. After these evenings out, Jacques Chemin and he liked nothing better than to walk home across Paris and hang out with the girls on the bridges.

Protestant Vacations

During the vacations, Yvonne de Gennes would send her son to Protestant families, in England in summertime, in the Alps in winter. He spent July 1948 in the family of the vicar of Bishopstone, a small village deep in the English countryside where he was afraid he might be bored. However, this did not happen, as the vicar also had two "*sassy*" young French girls staying with him, Monique and Sylvie: "*They would climb trees, they weren't afraid of anything. I didn't have their training, but my masculine pride made me follow them. We used to walk in the fields, help the farmers bring in the hay. We also played tennis.*

On one occasion, we played tennis on a Sunday, which caused a scandal. We got a very severe dressing-down: you just didn't play tennis on a Sunday."

He would spend his winter holidays skiing in the Alps. He stayed with Paul Faye, a veterinary surgeon in Bourg Saint-Maurice, recommended by a very straightlaced female friend of Yvonne de Gennes. *"That made me a bit wary. But Paul was very open-minded, very independent, an amazing discovery for me. He used to take me on his rounds. I can still see myself helping to castrate a horse in some remote mountain spot. I would hold its muzzle with a system to prevent it bellowing too loudly. And Paul used to take me skiing, with his chum, a primary teacher from Bourg Saint-Maurice. It was my first experience with good skiers. They were better than me, but I stuck with them, although I fell over a lot. We would sometimes indulge at lunchtime, and be a bit tipsy skiing in the afternoon. It was surreal".* He would also go to stay with the Morchs, a Protestant couple of around sixty who lived in Pierre-Grosse, an isolated hamlet 2000 metres up in the Queyras. *"In the morning, it was very cold in the house. I would go and wash in the fountain, which would be partially frozen. At dinner in the evening, a very frugal meal, they would say grace. Then Mr Morch would read a passage from the Bible",* he recalled. With them, he learned cross-country skiing. *"I would ski to the top of the mountain at five in the morning, getting back down around nine in the morning. I would be exhausted and Mrs Morch would say: "I've got just what you need" and bring me a bowl of hot salted water. It was probably good for hydration, but somewhat austere. It was a very Protestant atmosphere, which left a strong impression with me."*

"Sometimes I would go to Saint-Véran, a few kilometres away, where there was a dynamic young vicar I liked a lot and sometimes helped with his religious duties. On Sundays, he would hold services in tiny little mountain village churches for a dozen parishioners. We would go to the grocery in Saint-Véran and buy a bottle of wine, then to the bakers for a loaf. Then, in the church, the wine in the jug and the broken bread became something else. It was very beautiful. I don't know exactly why, but these unostentatious ceremonies with their extreme simplicity left a deep impression. They marked me and did much to make me a Protestant. With this pastor, I saw religion in a way that I hadn't suspected before", he explained. Previously, his mother had sent him to the temple in Auteuil, but he had shown a frivolous

interest in religious instruction. *"To prepare for confirmation, we had to read the Old Testament, and one morning Pierre-Gilles arrived all excited, announcing "I've come across a passage in which a man sleeps with his half sister",*[3] recalls Bruno Schroeder.

Baccalaureate

The school years were coming to an end. Pierre-Gilles was relaxed about the forthcoming baccalaureate examination. He gave the (false) impression of not working, but in fact he did not treat the exam as a formality. *"I remember revising a lot for history"*, he observed. And indeed, it was in the history orals that he experienced his only moment of stress. He was following Jean Maheu,[4] whom the examiner asked: *"Tell me about the Terror."* Jean Maheu began: *"The Terror was proclaimed following the dethronement of Louis XVI, on August 10, 1792, then…".* His exposition was masterly and left Pierre-Gilles gasping. He was afraid that his own effort would seem dull in comparison. In fact, he passed with an A grade. It was his first official success. After the announcement of the results (of the first baccalaureate), he wrote to his friend Bernard Prugnat:

"My dear Bernard,

I was sorry not to be there when you came the other day; I spent my last afternoon before the oral in a good game of tennis against Maheu. I passed this evening with an A grade (I know that you came close and undoubtedly deserved it as much as me). I owe it: a) to the debonair appearance of the teachers, b) to the luck I had in history-geography, where I was preceded by a herd of dunces, and profited from the comparison. The subjects: "Robespierre" (I had read a whole book by Mathiez about him) and "Comparative future of

---

[3] David's son Amnon, deeply attracted by his half sister, Tamar, yields to his desire and rapes her. Absalom, Tamar's brother and half brother to Amnon, murders Amnon two years later.

[4] Jean Maheu became President of the Georges-Pompidou Centre and then CEO of Radio France.

Algeria and Morocco" (hard, but interesting). In French, preparation in Marot and comprehension in Pascal (fascinating). In Latin: Livy (childs-play). Maths: a curious trigonometric property (I will show you it in October). English: Huxley's philosophy (a charming female examiner). Physics: paral-lel-sided blades and acids. Time to go, holidays tomorrow. Have fun by the Loire. I was sorry to hear about Vignal's failure, poor chap."

The serious stuff would start in the following year, in *classe pré-paratoire*, where students prepared for the competitive examination for entry to one of France's elite *grandes écoles*. His mother had dreamed of him becoming a diplomat. *"She saw it as the best job in the world. That was probably true when she was young. An ambassador had to take a huge number of decisions, without referring them to the Ministry of Foreign Affairs, but that is no longer true these days".* However, she made no attempt to interfere in his choice. On the evening of the bac-calaureate results, it was party time. Light of heart, he met up with Jacques Chemin and his friends for a night of jazz and jollity.

Chapter 3

Studious Years

Pierre-Gilles de Gennes had worked hard at lycée, but this was nothing compared to what awaited him in *classe préparatoire*... He would live the next two years like a long tunnel, only emerging for short intervals of light in the house his mother had just bought in the Southern Alps, where he could feel at home, recharging his batteries on mountain walks.

"In 1948, my mother had decided to find some little spot in the mountains where she could breathe the "fresh air"! She bought an old building near Orcières, a former farmhouse, with a sheep stable and a hay barn. She gradually refurbished it and it became the family haven, where we still stay today, with my children and with friends", he recounted. At the beginning, conditions were very basic. There was no running water. There was just one stove to heat the house, and the toilet was a wooden shack in the garden. Gradually, however, the house became comfortable, to the point that Yvonne de Gennes would divide her time between Paris and Orcières, living, as before, on her investments. *"Having been a nurse during the Great War, she fitted easily into the region. Whenever she went to anyone's house, she could always find someone of her generation to share memories with. People treated her with respect."* Especially as she created a library in the village and would visit old people: *"The Countess"*, as they called her, was a prominent local figure.

This house in Orcières became Pierre-Gilles' refuge from the age of 16, and the Champsaur Valley in which it stands, his playground. *"I would spend days on end walking in the mountains. I got to love hiking and, to a lesser degree, rock climbing. I would hunt marmots with a rifle, and we would eat them when we were camping up by the lakes. But you have to prepare the meat in a special way: first we would remove the skin, which we would give to a furrier in Gap; then we would carefully remove the fat, because marmots have a thick layer that smells disgusting; then we would leave the meat in a stream for about two days, to get rid of the smell. In the end, cooked with wine, it tastes like very strong game".*

He often invited his classmates to Orcières. His lycée companions, Bernard Prugnat, Boris Bespaloff and Jacques Chemin came there in the summer of 1949, after baccalaureate. The four of them would set off haphazardly in their boots in the early morning, to taste the joys of rock climbing. They had no pitons or karabiners, and would tie ropes around their waists for safety, *"So that, if one fell, he would pull all the others after him"*, as Jacques Chemin points out. They were fearless and would try every route until they reached the summit, frequently taking foolish risks. *"We didn't realise the danger, or rather, we didn't care. We came close to disaster several times. It still makes me shudder to think of it"*, he confesses.

Back at the house, the boys would sit down to eat, often with friends of Yvonne de Gennes, for example Monir and Madjid, a Persian couple that Pierre-Gilles would visit in Iran in 1973, and Nietta, his grandfather's second wife. *"My mother also still employed au pair girls, which gave the house a very cheerful atmosphere. We had a lot of them, Dutch or German... They often ended up in the arms of one or other of us"*, smiled Pierre-Gilles de Gennes. Dinner was a simple meal — a dish of pasta or lentils — but Yvonne de Gennes played her role as mistress of the house and presided over the table with regal authority. After dinner, they would all sit around the fire and conversation would flow. Nietta, a bookbinder of note between the 1920s and 1940s, would describe her encounters with André Gide, Jean Paulhan and Jacques Rivière, writers of the *New French Review*. Pierre-Gilles would sometimes get out his guitar and play a few fado tunes, or Monir might sing a Central Asian folk song or two. The boys would chat and joke together. They got on well, but essentially had little in common apart from their friendship with Pierre-Gilles, so that summer was the last time they would all spend time together.

An Unusual Choice

As the new academic year neared, Pierre-Gilles shut himself in his room and immersed himself in books on maths and physics. He would come back from his walks with frogs to dissect *"for practice"*. Indeed, he had a tough programme in natural sciences ahead of him, as he had not opted for a traditional *classe préparatoire* (maths and physics) — *"He was far too much of a maverick to stick to the beaten path"*, according to Jacques Chemin — but for a curriculum called *Normale sciences expérimentales* (or "NSE"), which included biology, with maths placed on the same footing as the other subjects. *"Because the selection was not based on mathematics, this class was more diverse and allowed other qualities, such as observational skills, to be expressed and recognised"*, recalled Pierre-Gilles de Gennes. Later, when asked about the reforms he would like to see in education, he would always cite NSE as an example. The idea of applying to this class had been suggested to his mother by the secretary at lycée Claude Bernard, Madame Nicknecker. *"It was good advice! She was probably also aware that the course had excellent teachers"*, he guessed. Their friend's choice came as no surprise to his classmates: NSE was there to prepare students for the entrance examination to the *École Normale Supérieure*, the holy of holies of France's elite schools, and none of them could imagine him heading for the *École Polytechnique* or some other engineering school, to follow a career in industry.

NSE was a stepping stone into the *École Normale Supérieure*, but only the best would be admitted. However, all the pupils in Pierre-Gilles's class had obtained at least a B+ grade in the baccalaureate — a *sine qua non* for entry into *classe préparatoire* at the Saint-Louis lycée — or an A grade. From the start of the academic year, the teachers' message was clear: *"There are 30 of you in the class. Last year, five dropped out and three ended up in psychiatric care. In the end, only five of you will get into the École Normale!"* Nonetheless, the atmosphere in the class was not bad. *"There were no vicious rivalries"*, recalled Pierre-Gilles de Gennes. The first weeks were tough, and...so were the ones that followed. *"The lycée was a dismal place. When I went back there almost 60 years later to give a lecture, it had been renovated and I didn't find it as dark and grimy as I remembered."* The regime was one of iron discipline. Anyone who arrived late in the morning was likely to find the door locked, and students had to form ranks before filing into class. *"We weren't allowed to go out. We were given detention every week and we were constantly hauled over the coals."* No one complained, everyone had just one goal: to pass the exam into École Normale. Pierre-Gilles, like the others, received his fair share of insults. *"In biology especially, my grades were average. When the teacher gave us a specimen to look at, my powers of observation were not very good, whereas some students had a real gift for it. You could say I was too much of a theorist. Even in maths, I remember getting a telling-off. We had a fearsome but excellent maths teacher called Maurice Chazal. "Even a clog-maker would be able to solve this simple little problem, and you can't!", he once shouted at me"*, he chuckled. *"The slightest misdemeanour led to 24 hours of detention, a whole weekend."* The day the maths teacher caught him reading a biology textbook in mid-lesson, the punishment was unequivocal: 24 hours detention. He landed the same punishment in physics. *"I was busy with, let us say, "nonscientific" subjects, and I failed to hand in my assignments twice in a row. I imagined that the teacher, Mr Legris — whom we actually liked a lot, he was always very calm — didn't keep an accurate account of the assignments handed in. Big mistake! I can still see him today, sitting at his desk, impassive: "De Gennes, two assignments missed: 24 hours detention." He was right to set me on the straight and narrow. It was essential at that stage to concentrate on work rather than chasing skirts"*.

The days followed each other, one much like the next. Pierre-Gilles would wake up early, set out from Rue Fantin-Latour and hurry sleepily along the banks of the Seine to catch the Metro. Once in class, he would nod off in the first hour when he had worked late the night before. He focused his efforts on the natural sciences, forcing himself to fill entire notebooks with sketches to sharpen his powers of observation, because his grades in the other subjects were good, although not as good as the best student in the class, André Authier (who would become a professor at Pierre et Marie Curie University in Paris, specialising in crystallography). However, Pierre-Gilles stood out from the other students, in this class too, for his facility and ingenuity. *"At the time, inspectors used to come round the classes. In mathematics, one of them had sent a student to the board to solve a problem. It was Pierre-Gilles, who had begun the calculations but was interrupted by the inspector: "No, that's not the way", he said, before starting on a long demonstration. At the end, Pierre-Gilles observed: "Look, Sir, I was on the same track." He had taken a shortcut and not argued: he let the inspectors finish"*, reported Jacques Signoret, a classmate and future professor of embryology in Caen. This incident gave Pierre-Gilles a certain kudos with his fellow students. But his popularity really took off the day the physics teacher set him the following problem: *"Two bubbles, one small and one big, are stuck together and separated by a thin partition. What happens when the partition is breached? Answer without thinking."* Pierre-Gilles calmly replied: *"That's exactly when you need to think."*[1] Teacher and students alike burst out laughing. His repartee hit home and became the class motto, repeated every time they were faced with a hard problem.

Pierre-Gilles swotted, showing a huge capacity for work and concentration, absorbing the toughest courses and storing up the essential fundamentals of maths and physics. *"Some people have appalling memories of their* classe préparatoire, *but I don't. It was hard, but ultimately healthy. In retrospect, I tell myself that the constant pressure was a good thing, because it kept us on our toes."*

He allowed himself little free time, but sometimes, on his way home after class, he would take a detour to a small shop in the Latin Quarter to listen to and buy jazz records. He was also developing an interest in classical music. *"I discovered classical music as a result of Cocteau's film* Les enfants terribles, *whose original soundtrack, especially a long Bach concerto, I found extraordinary! My friend Jacques*

---

[1] Contrary to what one might imagine, the small bubble deflates into the larger one, not the opposite, because being smaller, the pressure inside it is greater.

Signoret also introduced me to baroque music, as well as opera, which we previously found slightly ridiculous and old-fashioned," he recounted. *"Pierre-Gilles was curious about everything and very cultivated. One day, when our family doctor — Paul Viard, a very unusual man and a great lover of painting (he had treated the biggest names in interwar French painting, who had paid him in pictures, so that his dining room was lined with works by Modigliani, Fernand Léger and others) — was a guest in our house, Pierre-Gilles impressed him so much that our doctor invited him to dinner the following week",* recalled Jacques Signoret. The two young men would also attend occasional concerts and lectures on music.

However, Pierre-Gilles' first love was jazz and he would spend *"most Saturday evenings"* in clubs like the *Vieux colombier*, the *Rose rouge* or the *Tabou*, high spots of Saint-Germain-des-Prés, in company with Jacques Chemin and his clique. In the cellars of Saint-Germain, when everyone else was holding forth, Pierre-Gilles would stand slightly apart, cigarette in hand, rarely intervening, but always ready with a well-placed comment or witticism to set people laughing. One evening, he arrived with his new girlfriend, Nadine. *"She quickly fitted into my group of jazz lovers, which wasn't easy, because it was quite a closed group",* he recalled with pleasure.

Seaside Course

In the second year of *classe préparatoire*, the Easter holidays offered him a breath of fresh air before the entry exams. He undertook a course at the Banyuls-sur-mer oceanographic observatory, with a classmate Pierre Favard. *"We were welcomed by the director, Georges Petit, a specialist on cetaceans, a picturesque character with a good sense of humour, but we only had eyes for his daughter, Jacqueline, a very attractive girl with a Nefertiti profile. She impressed us by catching dogfish with her hands in the aquarium, since they are little sharks with a nasty bite. We became inseparable for the whole course",* he recalled.

It was the first time that Pierre-Gilles had set foot in a research lab. *"Aside from the coursework, I really enjoyed collecting calcareous algae to look at under the microscope. We also used to go into mountain caves to bring back strange cave-dwelling fish, which had lost their eyes in the darkness and developed other senses. It was very interesting. Jacqueline, Pierre and I would also go birdwatching in the wetlands on the edge of the Pyrenees. It was during that course that I really learned to observe. It only lasted two weeks, but it left its mark.*" He came back with dozens of sketches, but would sadly lose his sketch folder at the *Vieux Colombier, "a tragic loss",* in his opinion.

After these two idyllic and formative weeks, it was time to return to Paris... and start revising for the entrance exams. Predictions were rife. André Authier was the favourite. *"But I knew Pierre-Gilles well, and predicted that he would win, and that's what happened! His relaxed appearance disguised — with a certain disingenuousness — a huge capacity for work and he put in the required surge at just the right moment,"* reported André Authier. As the exams began, Pierre-Gilles retained his customary *sang froid,* as if the prevailing excitement had nothing to do with him, even arriving at one exam carrying his swimming gear. *"I also remember the botany exam, where I arrived firstly in shorts and secondly late. The teacher, Lucien Plantefol, an illustrious botanist, gave me a huge telling-off, but I still got through",* he laughed. *"For the oral on animal engineering, my examiner was Marcel Prenant, then a big name in biology, who was very nice to me. He showed me an object and asked me what it was. I had no idea... I was completely stumped! In fact, it was a wasps nest."* On the other hand, the mathematics oral gave him no trouble. Questioned by Gustave Choquet, a renowned mathematician who gave his name to the "Choquet simplex", he completed all the exercises, as he said, *"without help".* His examiner was impressed and asked him what he planned to do later on. *"After I told him that I wanted to do physics, he said: "Make sure you don't drop maths." It was good advice, which I only partially followed over the following years. But I caught up later, for example by teaching myself group theory all on my own",* he noted.

Top of the Class

The results of the entry examination to the *École Normale Supérieure* were announced in July, in Rue d'Ulm, and Pierre-Gilles came first in the NSE section. After the announcement, the students were given their traditional drenching. Whilst André Authier, who had also been admitted, prudently withdrew, Pierre-Gilles joined in the game, soaked but happy. He had no idea what the future held, or what he would do later, apart from *"physics"*, but he didn't care. The following year he would be entering the prestigious school in the Rue d'Ulm.

"My mother was pleased that I came first in the competition, but it wasn't a surprise: she took it for granted." Nonetheless, she proudly told everyone that her son had been admitted to the École Normale *"with the highest number of points ever achieved in the exam…"*, omitting to specify: in the NSE section. Lucien de Gennes, Pierre-Gilles' physician uncle, congratulated him and gave him a small motorcycle as a reward. Nothing could have pleased him more. *"That motorbike gave me a fantastic sense of freedom!"*, he recalled. Pierre-Gilles immediately took advantage of it: he bought a youth hostel map and set off *"pretty much on a whim"* towards the south of France. On the way, he spent a few days in the smart hostel in Villeneuve-lez-Avignon, resting and contemplating the Palace of the Popes, stretched out on the terrace with a book in his hand. *"Then I arrived in the Midi, but there were too many people so I set off towards the mountains. And there I had a wonderful time in the Guillestre youth hostel, at the edge of the Queyras in the Southern Alps: firstly, I met some wonderful youth hostellers — a female former dancer and a male photographer — with whom I stayed friends until their deaths, and secondly a young woman called Mathilde, who I fell deeply in love with"*. She was a little older than him, very beautiful and athletic. They went hiking and slept under the stars, spending an idyllic month together. She was his first real love, replacing all his former sweethearts. *"Sadly, we only saw each other once in Paris. I was fond of her, but my priorities had changed: I had just taken the plunge into the École normale"*, he confessed.

Chapter 4

Budding Physicist

With its worn floor and shabby walls, the main first-year student dormitory at the École Normale Supérieure was dismal. As the class of 1951 — some fifty students, just under half in the sciences, the rest in the humanities — took possession, a terrible hubbub echoed from the dingy ceiling, but Pierre-Gilles did not care: nothing could dim his pleasure as he stashed his suitcase in his "box" and amused himself by bouncing up and down on the bed. The fact that he had just joined France's most prestigious seat of learning was almost a minor detail, so happy was he to escape, at last, from his mother's clutches...

A new school intake means hazing! And the École Normale Supérieure might exist to educate the "nation's elites", but it was no exception. As soon as the semester began, the style of hazing set the base note for the irreverent spirit that reigned in the school. Lined up in Indian file, the conscripts (as freshmen were called), had first to bow, one by one, before the "Mega". This was the skeleton of a *Megatherium*, an imposing mammal that became extinct 10,000 years ago. It stood in the library, its attributes symbolised for the occasion by... a pair of slippers. This ritual accomplished, the treasure hunt began. All the students had a mission to fulfil or a "treasure" to bring back. The mission assigned to conscript de Gennes was to have tea in one of Paris' smartest *pâtisseries*, on rue de Rivoli, dressed as a boy scout. Pierre-Gilles therefore attired himself in overlarge shorts and a belt hung with bells and mess tins, which clanged together at every step. He was accompanied by a humanities conscript dressed as a priest.[1] As soon as the two customers entered the pâtisserie, they were the focus of every eye. The youngsters held their ground, soberly ordered tea and had themselves photographed. Then, delighted with the joke, they went back to the school where all the students were admiring the trophies they had captured: pedestrian crossing markers, street names, etc. There were enough objects to open a bazaar! Two conscripts had even persuaded Jean-Paul Sartre to dedicate his book *Being and Nothingness* to the christian philosopher, Gabriel Marcel. The existentialist writer had written: *"For you, Gabriel, this essay on being so that you can respond from the perspective of nothingness"*, the irony of which triggered thunderous applause. In the evening, the conscripts underwent a final challenge: they had to run around the Panthéon with a sheet over their heads, moaning like ghosts. It was then that the revellers realised that two conscripts were missing, Philippe Nozières and Jean-Pierre Beaufils. Enquiries were instituted. In fact, they had been caught removing a store sign for the treasure hunt, arrested and kept in jail overnight! The hazing finally ended with a big reconciliation drink. *"I have opposed hazing, which I am convinced can be dangerous* (Author's Note: he would refer to the *"long slavery, damaging both to personal life and education"*, in his preface to the book on the subject by Aude Wacziarg[2]), *but the version we experienced at the École Normale was very mild and left no one with any traumatic memories"*, explained Pierre-Gilles de Gennes.

---

[1] Paul Veyne, a French historian specialising on ancient Rome. Pierre-Gilles de Gennes and he would find themselves together again at the Collège de France, where Paul Veyne was appointed four years after the physicist to the chair of Roman history.

[2] *Bizut. De l'humiliation dans les grandes écoles*, Austral, 1995.

Hellraisers

A small band of merry men quickly formed amongst the science students, consisting of Philippe Nozières, Roland Omnès, Jean-Pierre Beaufils, Jean-Paul Bloch, Pierre-Gilles, along with two friends who had been in *classe préparatoire* with him, Pierre Favard and Maxime Guinnebault — so inseparable that they were called Favault and Guinnebard — and finally, Pierre Averbuch, a pillar of the Communist Party, who would join them when not busy distributing tracts or handing out petitions[3]... On Sunday mornings, this merry band, still in dressing gown and slippers, would cheerfully meet up in a cafe near the school, for example opposite the church of Saint-Jacques du Haut-Pas, to eat breakfast and contemplate *"the pretty parishioners emerging from mass"*. From time to time, they would conduct "punitive expeditions" against students judged to be bores or a pain in the neck. *"One-day, we daubed a student with different coloured paint — it was the kind of thing that was common in the school — and told him that, to clean him off, we were going to throw him into the Ernest pool"* (Author's Note: so called because of the fish introduced into it by Ernest Bersot, former director of the school). *He didn't think that we would do it. Two of us spontaneously grabbed him by the pool and and threw him in the water. I can still remember his face when he realised that we were going to go through with it. It was a bit cruel, but not excessively, as the water wasn't very cold..."*, chuckled Pierre-Gilles de Gennes.

The little band also made its mark outside the school. *"One day, Maxime Guinnebault arrived in triumph with a ticket for a popular operetta that was playing at the Gaîté Lyrique,* Les cloches de Corneville, *by*

---

[3] Philippe Nozières would become a professor at the Collège de France, Roland Omnès emeritus professor and President of South Paris University and Pierre Averbuch Director of Research at the Very Low Temperature Research Centre in Grenoble. Jean-Pierre Beaufils would be appointed director of the Central Chemistry Laboratory in Lille, whilst Jean-Paul Bloch, a class ahead of the others, son of the historian Marc Bloch, would be put in charge of the science laboratories of the French Southern and Antarctic Territories. Pierre Favard would become Director of the Experimental Cytology Centre and a specialist in microscopy (there is a prize named after him).

Robert Planquette. We tried to trade his seat in the stalls for seven seats in the gods, but it didn't work and we ended up paying for them ourselves", remembers Philippe Nozières. With their scruffy clothes, they drew stares as soon as they came into these plush surroundings, especially Pierre-Gilles, *"the most hobo looking of all of us"*, with his shorts and long hair (very rare in the 1950s). As one scene followed another, the lads applauded ever more loudly. When the best-known tune — *"Ding, ding, ding, ding, ding, dong. Ring, ring, ring, joyful bells!"* — came to an end, they applauded even louder, with fake enthusiasm. *"Encore, encore! It's wonderful!"*, cried Roland Omnès. Flattered, the singer began again, sparking hilarity in the group. One member of the audience, realising that they were taking the micky, shouted: *"Gentlemen, we are not here to have fun"*; Pierre Favard drew himself up and haughtily replied: *"So I suppose, sire, that you are here to think?"* The gang's adventures and escapades were legion!

Between these *manips* (as such capers were called in the jargon) and work, the students scarcely left the school. *"We would sometimes meet around a gigantic cheeseboard in one of the bistrots on Rue de l'Épée de bois, and tell stories of all kinds. It was great fun"*, recalled Pierre-Gilles de Gennes. *"We would go to the movies from time to time, but we didn't have much money..."*, adds Philippe Nozières. The school was their whole life: they ate together in the *pot* (the refectory) morning, noon and evening, worked together in the *turnes* (study rooms for four or five students) and slept in the dormitory (they were only entitled to their own room after the second year). On warm nights, they would get together on the roof of the school, to talk about anything and everything, to joke about the latest pranks, before abseiling down a rope to the lower floors.

Free Electron

Pierre-Gilles, however, was something of an exception in the group. Unlike his companions, he did not spend all his days within the walls of the school. *"We didn't often see him at the school"*, confirms Pierre Averbuch. He would come and go as he wished. *"He always got around by motorbike, on which he had an interesting theory: he used to say that since junctions were very dangerous, one should spend as little time as possible on them and cross them at full throttle, which is what he did"*, smiles Philippe Nozières. No one knew where he went, or what he did, but it was enough in his second year for a blonde to come out of his room one day and a brunette the next, for the students to credit him with a multitude of conquests. *"When you go to see de Gennes, make sure you knock before going into his room"* was a standing joke in the school. However, no one knew anything about his life, because he did not take people into his confidence, even friends.

True, there were affairs of the heart — *"Not all of them merit a detailed description"*, he smiled, *"Nevertheless, I am somewhat sentimental and I also had a lot of women friends"* — but despite that, he spent most of his time on his work. *"Ah yes, there were classes"*, recalls Roland Omnès. *"In theory, we were supposed to go to the university, but we found some of the lectures so bad that no one went, apart from one or two who would take on the job of telling the others what they were about."*

It is true that, since the First World War and the decimation of whole cohorts of students, the university had declined. The situation was exacerbated when the Second World War isolated France from developments in international research, to the extent that the physics taught at the Sorbonne in the 1950s was completely out-moded. There were no classes on statistical physics or quantum physics. Students at the École Normale would tell mocking stories about the ignorance of certain teachers. *"There is a more modern presentation of what I am explaining, Maxwell's equations — but it is too complicated, you will come to that later"*, one electricity teacher had told them, at a time when Maxwell's equations were already more than 80 years old.

Pierre-Gilles wanted to form his own opinion and put together his own personal timetable of classes to attend, although one of the criteria was that they should not be too early in the morning. So he skipped the classes on probability (which began at 8 a.m.), but attended Georges Valiron's classes on differential and integral calculus, and those of André Lichnerowicz on mathematical methods for physics. He also rated Henri Brusset's course on mineral chemistry. *"He had an armada of little helpers around him, busy running the experiment he needed to illustrate his class. I have good memories of him."* We also know that he was a regular at the chemistry classes given by Paul Pascal, an old teacher who looked like a relic of the previous century. *"He was already an elderly gentleman who had retained the style of a previous era with his handlebar moustache, starched wing collars and neat little bowtie. He passed on a number of ideas in physio-chemistry which I still use today, for example ways to understand the viscosity of a gas, the magnetism of transition elements... Simple, but very useful notions."* Unlike some of his fellow students, who denigrated chemistry — considered as less noble than physics in the ranking of the sciences established by Auguste Comte, with mathematics at the top, followed by physics, then chemistry behind — Pierre-Gilles was interested in chemistry to the point of having had private lessons the year before. *"I had a few lessons on mineral chemistry in* classe préparatoire *at the home of a young lecturer called Faucher — a huge fellow, taller than me! — who taught me the chemistry of solutions and got me to do experiments, so that when I arrived at the École normale I had a more complete picture of chemistry than the average freshman".* In his class schedule, Pierre-Gilles also chose to include lectures given outside the Sorbonne, for examples those by Yvette Cauchois on spectroscopy or by... his mentor, Edmond Bauer, at the Institut Curie. *"I used to go and listen to Edmond Bauer, who gave a short introductory course on statistical and quantum physics. I think I was the only École Normale student to attend them"*, he suggested.

At the École Normale, he would also pick and choose his classes. He gave up going to maths courses, even those given by Henri Cartan, co-founder of the Bourbaki group (*"It wasn't my cup of tea"*, he confessed). *"In any case, there weren't many maths classes that we, as physicists, had to attend. In the end, we acquired mathematics, the tools, through quantum mechanics and statistical physics."* By contrast, he never missed a single physics class at the school, where all the teachers had, in one way or another, left their mark on the history of the discipline. *"Alfred Kastler's classes* (Author's note: Nobel Prize for physics in 1966 for the discovery of optical "pumping", one of the keys to the discovery of lasers) *were excellent. He had a gift for teaching, managing to be effortlessly simple. He was also a very nice man"*, recalled Pierre-Gilles de Gennes. He also attended the classes given by Yves

Rocard, the director of the physics lab, whom the students thought very highly of (it should be said that he gave them his complete trust, to the point of telling anyone who asked him for a reference: *"Write it yourself, and I'll sign it."*). He was an unusual kind of scientist and a pioneer of interdisciplinarity. *"Having done research both on instabilities in bridges and on radio lamps, he was known in the fields of mechanics and electronics, which were very different worlds"*, recalled Pierre-Gilles de Gennes. Yves Rocard was also a keen defender of scientific applications and no believer in "theory for its own sake". His classes reflected this approach: he explored different aspects of classical physics in terms of concrete cases, explaining, for example, how nozzles convert the energy of burning gases into kinetic energy or how car suspension systems should be made in such a way that their vibrations are synchronised with human steps. Throughout his education, Pierre-Gilles would see Yves Rocard as one of his intellectual models, appreciating the way his teaching was *"anchored in reality"*. Although clearly gifted in theoretical aspects, the young man's interest was already manifestly focused on practical physics.

By contrast, he left off attending the classes given by Louis de Broglie, the leading light in quantum mechanics in France in the 1920s, *"a very distinguished elderly gentleman"*, he recalled, but one who was unfortunately content to read his notes at the board in an inaudible voice, inevitably losing his students' attention. In the 1950s, quantum mechanics was not yet taught anywhere else (or was badly taught). So Pierre-Gilles and his fellow students would soon be forced to take the bull by the horns...

Self-Taught

Once he had fitted these classes into his timetable, the young man was left with long stretches of free time, which he used to devour physics books by the dozen. He spent hours browsing through the shelves of the science bookshop on Rue Gay-Lussac, reading textbook after textbook to flesh out the knowledge acquired in class... or to fill in the gaps. In fact, it was from books that he taught himself the basics of quantum mechanics. In the absence of "good" classes, there were fortunately "good" books, some of which had just been published, such as Leonard Schiff's *Quantum mechanics* in 1949, or Paul Dirac's slightly older *The Principles of Quantum Mechanics*, the "bible" that reconciled the algebraic approach of Heisenberg's matrices with Schrödinger's wave function approach. *"Pierre-Gilles had decided to read the Dirac. He sat down in his room one evening and read it in a single night, with a bottle of cognac and cigarettes beside him"*, recounts Roland Omnès. *"The next morning, we said to ourselves that it wasn't a bad idea..."*, so Pierre-Gilles' unorthodox method attracted imitators. In the weeks that followed, the counterintuitive concepts of quantum mechanics became the subject of endless discussions in the cafes... *"We were fascinated by science and especially by quantum physics. But we weren't just fascinated: we also knew that it was important and that we needed to understand"*, explains Roland Omnès. Young as they were, they were aware that quantum mechanics was the foundation of whole swathes of modern physics and that it was therefore an essential tool for a would-be physicist.

Now Pierre-Gilles wanted to go further with the ideas he had acquired, but to do that he had to master group theory. *"But we weren't being taught any of that kind of algebra, it was a complete blank! So we had to learn everything for ourselves"*, he explained. So he unearthed a book that outlined the bases of the theory, unfortunately written in German. No matter, he got hold of a German language book and learned the rudiments he needed to decipher the mathematics manual! Nothing would stop him and that is how he worked all his life, *"equipping himself as he went along"*, i.e. progressively learning what he would need from books, using the method that began in his sabbatical year.

The young man did not confine his reading to textbooks, but from 1953 began to read scientific articles. He started by looking up the odd reference in bibliographies, and gradually broadened his reading, eventually coming to Richard Feynman's articles on superfluid helium: a revelation! *"Feynman changed my entire perspective. God knows he discovered loads of stuff on the algebra of operators — the high wire act of mathematics — but the really big deal was his understanding of physics. I realised that a knowledge of Navier-Stokes equations*

(Author's note: the fundamental equations of fluid mechanics) *doesn't mean that you can understand hydrodynamics"*, he explained. It was at this point that the student threw off the straitjacket of the *"young academic theorist"* and, taking Richard Feynman as his scientific model, began to focus on understanding the physical meaning of phenomena. In practice, he forced himself to think before plunging headfirst into equations, to reject formalism in favour of understanding, and in so doing achieved his first successes. *"I remember one occasion that gave me immense satisfaction. Rocard had set us a problem, which wasn't very hard when I think about it now. Roland Omnès had found an answer that was correct, but was based on a rather complicated approach, whilst I found a simpler method. It was something about spinning electric motors and I had introduced a time-dependent parameter, which showed what was going on in a few lines. Rocard said: "Omnès has found a solution that is strictly correct, but I prefer de Gennes' approach"*, he recalled. The teacher's comment was superfluous, because all the students were left openmouthed by their colleague's exposition. For the first time, Pierre-Gilles had distinguished himself from his brilliant classmates at the École Normale. *"At the age of 20, he was already amazing. He had a maturity in physics that I only acquired later, after my doctorate. He also demonstrated a great economy of means and a very open mind"*, avers Philippe Nozières.

Political Awareness

Apart from the group of friends in his year, Pierre-Gilles barely socialised with the school's other students. *"I spent little time with the literary students, apart from a few such as Claude Nicolet, a specialist on ancient Rome, or Bernard Bourgeois, President of the French Philosophy Society, who used to talk to me about Leibniz's monads. They were real thinkers, whereas a lot of the literary types I met at the school seemed to me to be blinded by their ideals, using philosophy solely to justify their communist ideas. I was very suspicious of philosophers on the whole"*, he confessed.

In keeping with the climate of the time, the École Normale was highly politicised, and Pierre-Gilles went so far as to consider certain of its leading lights, such as Louis Althusser, Emmanuel Le Roy

Ladurie or Jean-Baptiste Gourinat, as *"gurus of a kind"*. So one
February evening in 1952, at a debate run at the school on biological weapons in Korea — the US was accused by the Chinese and
North Koreans of having used bacteriological weapons in the
Korean War — Pierre-Gilles pointed out that there was no evidence
for these claims. In response, the communists, convinced of the
guilt of the US, unanimously shrugged off the argument. *"My distrust
of philosophers was forged in situations like that"*. The divide with the
"literary set" continued to widen. On Stalin's death on March 5,
1953, complete silence reigned in the refectory, as if the communist
students were in shock. *"Pierre-Gilles seemed upset, as if wounded by this
reaction, which the rest of us simply saw as over the top"*, recalls Philippe
Nozières. *"One of the key moments of my education is still the day of Stalin's
death. The consternation of his many acolytes. Their words: "We haven't lost
everything, there are still his writings, etc." And 40 years on, what they have
become (stalwart bastions of bourgeois society). Lessons like that you don't forget"*, he wrote in 1994.[4] Having flirted with the idea of becoming a
writer just a few years before, he did not regret his decision, so little
did he feel he had in common with this set. After this, he would stay
at arm's length from their overly sectarian political debates. He also
refused, with just a few exceptions, to sign the hosts of petitions
going the rounds — the Stockholm Appeal against nuclear
weapons, execution of the Rosenbergs, etc. — and, whilst the other
members of his group would read the texts and pretend to discuss
them seriously he preferred to withdraw discreetly. Later, in particular after his Nobel prize, he would publicly commit to many causes,
but always with circumspection. *"I have rarely felt in tune with the commitments of French "intellectuals". The case of the Kravtchenko trial in 1949
is typical. He was a Russian engineer who defected to the West and wrote* I
Chose Freedom, *a critique of the Soviet system. The Communist Party
claimed that he was lying and many French communist intellectuals spoke
against him. Ten years later, everything Kravtchenko had said turned out to*

[4] Alain Peyrefitte, *Rue d'Ulm. Chroniques de la vie normalienne* (édition du bicente-
naire), Fayard, 1994.

be true. Broadly speaking, I find that the French intellectual community often espouses causes that I do not share. That is still true today. For example, the way French intellectuals approach ecological problems is not how we will succeed in solving them."

As the end of the year approached, revision began for the examinations, since students were required to obtain the necessary certificates to be awarded their end of year diplomas (bachelor's degree in the first year and master's in the second). *"We didn't worry much about those exams. The main thing was not to fail, so as not to spoil the vacation"*, recalls Roland Omnès. *"In fact, we started preparing for the exams just a few weeks before."* While the others doggedly revised their courses in general chemistry, Pierre-Gilles played the guitar in his room, as if he did not need to prepare. He always gave this impression of finding things easy, notably in his classes at the Collège de France, but make no mistake, while he was certainly immensely gifted, his apparent facility also came from relentless work, which often went unnoticed. He was not alone in showing this particular form of snobbery — particularly widespread in the scientific community — of pretending that everything comes naturally. In fact, however, even for him, not everything came easily: he passed his diploma, but failed the test on mathematical methods in physics, as did Philippe Nozières. *"We both went to the exam with our hands in our pockets, convinced that we would be able to do it all, which is typical behaviour for the École Normale. And we didn't get a single thing right. It was a valuable lesson. But we didn't have to redo the exam, as it was not compulsory for us"*, he explained. Perhaps his mind was also elsewhere, dreaming of his imminent departure for an unusual expedition?

Conquest of Iceland

At the beginning of July 1952, Pierre-Gilles boarded a ferry for Iceland. Over the year, he had lost interest in the natural sciences, but the moment his biologist friends, Pierre Favard and Maxime Guinnebault, told him of their plan to study the flora and geology of Iceland, he rediscovered his naturalist instincts and volunteered to join the trip, which also included Jacques Signoret (their NSE classmate now at university) and Jean-Paul Bloch. *"We set off with two tents, a bit of money and a huge, incredibly heavy camera! The crossing to Reykjavik was long, and we were all sick as dogs!"* Once ashore, they criss-crossed the island, crossing huge lava fields, climbing the Sneffels volcano, starting point for Jules Verne's *Journey to the Centre of the Earth,* and exploring the solfataras, those lunar expanses of volcanic vents with their hydrogen sulphide fumes. *"Unfortunately, there was also a lot of mist. We hadn't realised that it was very dangerous and, sud-denly, Jean-Paul Bloch stepped up to his knees in a kind of mud that must have been around 150 degrees. He got badly burned".* Bad luck pursued them. *"A sheet ate one of our tents, so we had to squeeze into the only remain-ing tent. In addition, we didn't have much to eat: it was physically pretty tough",* he recalled. Continuing eastwards, they arrived in a small port, enclosed by ice five months a year, where they were warmly wel-comed and put up in a municipal school room, which fortunately had a stove. They had not eaten anything hot for a week! Here, they made friends with a villager who had a car. *"He was very proud of it. We thought it was weird to have a car, as there were only a few kilometres of tarred roads around the village, and just rough tracks beyond that. We understood what he used it for when he took us for a ride into the hills, as far as the car could go. There, he opened a big trunk in which he kept slices of shark meat and, more importantly, bottles of home-made spirits! We drank his schnapps and chewed shark. A great memory!",* he laughed. After this, the companions continued their journey, still in difficult conditions. They visited the great glaciers of the south, crossing fast-flowing rivers with their luggage on their heads. *"I had an advantage compared with the others: being taller, I could cross without getting my shorts wet, whilst the others got soaked to the waist."* After a month, they returned to France, *"somewhat thinner",* but happy. They brought back a film in which Pierre-Gilles, starring in one sequence, dives into very blue water in a fault running across an old lava flow, then re-emerges fur-ther on and picks a flower on the bank. *"The film probably is probably still around somewhere…",* he commented, smiling in recollection of the cliched scene.

Summer School at les Houches

In second year, Pierre-Gilles moved into a double room, an opportunity for... his mother, still seeking the slightest pretext to stick her nose into her son's life! She took charge of the room's decor, choosing the fabric and having matching bedspreads and curtains made, under the young man's agonised gaze. He would rarely visit her, but for his part could never be sure not to bump into her on some corner of a school corridor. Indeed, she was quite ready to come into the building and move from floor to floor, bellowing out: *"Pierre-Gilles! Pierre-Gilles!"* Depending on his mood, he would either emerge or lock himself in his room, waiting for her to get fed up and leave. *"Even when we were no longer living under the same roof, she still had a tendency to tell me what to do. So we still had arguments, even though I had gained my independence."*

The second year went by without problems. In the spring, a young teacher at the school called Pierre Aigrain suggested to Pierre-Gilles and his physicist colleagues that they should attend the "Les Houches summer school for theoretical physics", which had been founded in 1951 by a 27-year-old theoretician, Cécile de Witt. She was doing physics research in the USA, where she had moved after marrying an American physicist, Bryce de Witt, giving up her position as a lecturer in Nancy. Aware how far behind France was in physics research, though now a US resident, she wanted to help close the gap. It was this that gave her the idea of creating a physics summer school in France, *"at a time when nothing like it existed"*, added Pierre-Gilles de Gennes. She spared no effort, knocking on every door to obtain funding, and succeeded in creating the summer school at Les Houches, in the Alps. *"The organisation was cobbled together, but she managed to attract the top names in physics, who were paid peanuts but were attracted by Cécile's enthusiasm"*, explains Philippe Nozières.

Pierre Aigrain went on to explain that there were only four places available for the whole class year. *"Who's interested?"* All the hands went up. To decide, he asked them to prepare a paper: the best would be selected. A few days later, Pierre-Gilles, somewhat sheepishly, gave a paper on the "Klein-Nishina regime", a particular case of Compton scattering, the collision between a photon and electron. He demonstrated the formula but admitted, with some embarrassment, that in some cases the equations had gone wrong. Pierre Aigrain smiled: *"Don't worry. The problem is not without difficulties!"* Pierre-Gilles was selected, with Philippe Nozières and two other students. The school would only begin in three months, but he immediately began to prepare.

When Pierre-Gilles arrived in Les Houches in early July and saw the chalets with their magnificent view over the Mont Blanc massif, he immediately saw himself hiking in the mountains. In fact, immersed in work, he had time for no more than a few short walks. After meeting the other thirty or so participants from elite European schools and universities, he took a tour of the premises. *"Comfort was minimal. All of us, students and teachers, were housed in small chalets, which got their water supply from springs. The classroom was also in a chalet, equipped with a blackboard and canvas chairs."* The classes, mostly in English, took place in the morning, six days a week. Afternoons were free, but not a time for twiddling your thumbs! *"The atmosphere was extraordinary. We had all the physics of the era right in front of us! Nozières and I worked day and night trying to decipher and restructure what we had heard during the day"*, he recalled enthusiastically. At the end of the day, the participants attended a seminar or, on free evenings, were trucked down to the village for dinner.

"I learned more in those two months than I would have done in a year at university. I discovered the whole of contemporary physics, quantum mechanics, statistical physics, meson physics…" One of the most striking speakers was Rudolf Peierls, a leading light of quantum mechanics in the 1930s, who covered topics in solid-state physics that the students had never heard of. *"In addition, he was very funny. Pierre-Gilles and I were staying in the stable above his chalet. In the evening, it was a 15 minute walk in the dark to get there. His Russian wife used to give us tea. They were a really nice family"*, remembers Philippe Nozières.

Another speaker who made a sensation was the American William Shockley from Bell Labs, who turned up in a red Jaguar convertible! The man who in 1947 invented the transistor, the basic component of microelectronics and computers, only stayed three days, but his classes marked Pierre-Gilles, both in form and content. *"We were very well-behaved students, and rarely, if ever, asked questions. That was the way in the French system"*, he explained. But William Shockley wasn't having any of it. *"As you're not asking any questions, I'll ask them for you"*, he would announce, before posing a series of searching questions about his own course. The young theorist discovered the effectiveness of interactive teaching, standard practice today, but not at the time.

However, the star of the Les Houches school that summer was unquestionably Wolfgang Pauli, one of the pioneers of quantum mechanics, who Einstein had said had understood relativity better than himself (aged 25, Wolfgang Pauli had formulated the exclusion principle that now bears his name, a fundamental postulate which states that two identical fermions (e.g. two electrons) cannot occupy the same quantum state simultaneously). The students were meeting a legend. Alas, his courses were abstruse. He never wiped the blackboard, which meant that at the end of the course it was completely covered in illegible chalk equations. The participants did not get much out of them, but did not dare to complain in front of the master, who was invariably serious. *"I remember that Shockley used to tell stories that made us all laugh, except Pauli, who would just show a hint of a smile*, recounted Pierre-Gilles de Gennes.

The stuff taught by these masters of 20th-century physics was of a very high level. *"Don't worry if you don't understand everything: learning happens by osmosis. You will understand in a few years"*, Cécile and Bryce de Witt would repeat, to encourage the students. Some more or less threw in the towel, but Pierre-Gilles and Philippe Nozières stuck to it, sometimes working far into the night. *"We really wanted to understand as much as possible"*, emphasised Pierre-Gilles de Gennes. They studied together and shared the same chalet. That summer brought them closer together. *"I really got to know Pierre-Gilles at that time. We became very good friends"*, recounts Philippe Nozières. Thus began a relationship that was to be a mixture of friendship and rivalry.

At the end of the two months, students could opt to take a test, choosing one subject from a dozen options, with 24 hours to prepare. Pierre-Gilles chose: *"Adiabatic processes in thermodynamics and quantum mechanics"*, proposed by Rudolf Peierls. No other students selected the subject, and with good reason! No one understood the title... The young physicist bent over backwards to build his talk around the different meanings of the word "adiabatic" (A4-1).[5] At the end of his presentation, Rudolf Peierls asked him a question on phase transitions. *"I completely dried up. I had nothing to say. So he said to me, very kindly: "Don't worry: it took Ehrenfest six months to answer that question* (Author's note: Paul Ehrenfest was one of the founders of quantum and statistical mechanics). *Congratulations on your presentation"*, he reported. *"I can no longer remember what Pierre-Gilles said that evening, but I remember that Peierls was amazed by his work"*, recounts Philippe Nozières.

Pierre-Gilles would often repeat that those two months were some of the most important of his life. *"I came back from that summer school a changed man: my vocation for physics was born there."* He would also emphasise the impact of the school beyond his own personal case. *"Our generation was saved by that school."* According to him, Cécile de Witt *"did as much for physics in France as many governments.... It is no exaggeration to say that [the] creation [of the Les Houches school] was one of the major factors in France's post-war scientific revival (...). I wanted to tell you how much, in my opinion, the nation owes her"*, he would write to the Director of Higher Education in 1988.

After these two exciting months, Pierre-Gilles regretted having chosen Edmond Bauer's laboratory at the Institut Curie for his third year internship. It had been agreed that he would study questions of molecular spectroscopy in microwave radiation (a technique designed to determine the composition of gases by radiation absorption), but he now found the subject somewhat uninteresting. So he decided, like Philippe Nozières, to do his internship in a leading-edge laboratory, at the École Normale. *"But I felt uncomfortable about telling Edmond Bauer that I wasn't coming. I went to see him at home in Saint-Gervais, near Les Houches, where he had a fine house. He was very understanding."*

---

[5] The appendix "Find out more" on page 320 provides additional scientific details.

The Freshmen's Ball

The lab where Pierre-Gilles took his first steps as a researcher was headed by... Pierre Aigrain, the same man who had told his cohort about the Les Houches school. He was a *"young and dynamic"* teacher, who had just returned from the US, where he had studied electronics during the war. *"Yves Rocard then gave him the job of creating a semiconductor lab at the École Normale and Pierre Aigrain thus founded the science of semiconductors in France"*, noted Pierre-Gilles de Gennes. Pierre Aigrain, as Philippe Nozières was to write in a tribute in 2002, was one of the *"physicists who left a profound mark on research in France since [1950]. He is one of the pioneers of solid-state physics and his work on semiconductors was behind the rise of modern electronics."*

"Pierre Aigrain turned up every morning with new ideas", recalled Pierre-Gilles de Gennes admiringly. *"He devised amazing effects for semiconductors, imagining what would happen with a current, with light or a magnetic field..."* Students would pluck his ideas out of the air and try to test them by mounting experiments, with equipment grabbed where they could find it, *"ideal conditions to stimulate a readiness to improvise"*, noted Pierre-Gilles de Gennes. Indeed, the laboratory had few resources. In winter, the temperature rarely rose above 17°C, which caused a misunderstanding when Pierre Aigrain compared measurements of the resistance of germanium at room temperature in his laboratory (70 ohm-cm) with those obtained at Bell Labs (47 ohm-cm): in fact, the disparity was caused by the difference in room temperature between the two institutions!

Pierre Aigrain gave the young man his first research topic. *"He thought that the action of certain catalysts could be explained through semi-conductor theory, and asked me to study magnesium oxide, a simple solid that was known to be a good catalyst. "Find us some crystals and make the appropriate measurements", he said. I began by going to see an old crystallo-grapher at the Natural History Museum, but didn't find what I wanted."* The only thing to do was make those dammed magnesium oxide crystals himself from magnesium powder. *"I remembered an article in*

Science et vic, *which I had read every month in high school, which described Verneuil's method for growing a crystal from amorphous powder over a flame*".[6] He ran the experiment, but every attempt failed. *"So I took the bull by the horns , found a graphite cylinder the size of a one litre bottle in one of the school attics and filled it with magnesium powder."* His idea was that, since graphite conducts electricity, running a current through the giant crucible would heat it by a simple Joule effect, melting the powder. Then all he would have to do was let it cool down to produce crystals. The experiment was simple, but required so much power that he had to go round all the labs warning them to switch off their experiments for fifteen minutes. The school was abuzz and everyone came to watch the experiment, though from a safe distance. Pierre-Gilles connected the apparatus. Everything was ready! *"We hid behind the furniture to watch the cannon, which turned... cherry red"*, he recounted. The experiment worked! *"But there was one problem I had not foreseen: the thermal gradient between the centre and edges of the cannon was so high that the crystals, instead of being straight, were curved like bows and... unusable"*, he sighed.

After this failure, Pierre Aigrain placed him under the supervision of a young female researcher and asked him to produce thin films of germanium (a metalloid with excellent semiconducting properties), not using the standard technique of chemical etching — in which the thickness of the resulting film cannot be controlled — but by testing a new type of electrochemical etching. The principle was as follows. While the germanium is thick, the electrical current can travel through it and the thickness of the layer diminishes, but when it reaches a certain value, the current can no longer travel through and the electrochemical reaction stops (A4-2). It was an ingenious idea, *"typical of Aigrain"*, commented Pierre-Gilles de Gennes, but difficult to implement. The young physicist and his scientific instructress worked for hours in the dark (light produces parasitical charge carriers) without making any good quality films.

---

[6] Auguste Verneuil discovered this flame fusion process in 1902, a method widely used subsequently to produce synthetic precious stones, for examples the rubies used in the first lasers.

The apprentice chemist handed back his white coat and returned to his office.

"I was failing again and again, but it didn't matter, because I was learning so much." These experiences taught him about the vagaries of the laboratory bench: he understood the chasm between a good idea and its realisation — a lesson that later on made it so easy for him to talk to experimentalists. The other interns also had their share of failures, even Philippe Nozières, with his reputation for dexterity (*"I have seen him completely disassemble and reassemble his washing machine"*, Pierre-Gilles de Gennes would say, admiringly). *"Nozières had moved heaven and earth to design and build a metal frame. When he wanted to bring it into the lab, the frame was too big to get through the door. In a rage, Nozières started sawing it into pieces"*, he remembered, still laughing.

Giving up the idea of getting him to do experiments, Pierre Aigrain suggested that Pierre-Gilles should present a seminar on *"a new effect described in* Physical Review: *the Overhauser effect."* The young man agreed and spent a fortnight swotting. On the day, he began his presentation in front of an audience made up of students and researchers from all over Paris. Despite this, there was no stage fright: he had worked hard enough to feel confident of his subject. *"But while I was talking, a little man with a moustache kept on asking me questions, of a fairly technical kind, which I did my best to answer."* At the end of the presentation, he was wondering who this "pest" was and Pierre Aigrain introduced him to Anatole Abragam, a researcher at the AEC (Atomic Energy Commission), equally well-known for his work on nuclear magnetic resonance as for his killer jabs. The young man was wary, but the AEC physicist congratulated him warmly. In private, Anatole Abragam commented to Pierre Aigrain: *"That lanky young fellow will go far."*

The end of the year was approaching. *"I learned a lot from Pierre Aigrain, not only about semiconductor physics, but also a particular way of working, searching out ideas and orders of magnitude to assess different physical effects. For example, Pierre Aigrain always encouraged us to refer to a book that we called the Bible, a sort of directory which listed all the standard properties of materials, such as the refractive index, thermal conductivity, etc.*

A good habit." After this, the young physicist would always juggle orders of magnitude when tackling any kind of problem, to get an idea of the result before starting his calculations.

Pierre Aigrain was the first researcher the young physicist spent time with, and there was nothing ordinary about him. "*He had a very different style from other researchers. He was brilliant and not at all pretentious, which is quite rare.*" He was also so passionate about science that he neglected to publicise his discoveries. "*He was the man who invented semiconductor lasers, but he didn't publish. He didn't like writing, and gave his ideas away freely*", explains Philippe Nozières. He also treated his students as colleagues, "*although we were really young*", to the extent of taking them to the International Semiconductor Conference in Amsterdam, "*which no one did in those days*", commented Pierre-Gilles de Gennes. Very excited, they squeezed into a big late model Frégate Renault, and Pierre-Gilles attended his first international conference. Finally, Pierre Aigrain was a theorist who listened to industrialists. All these were characteristics shown by Pierre-Gilles de Gennes, once he became a researcher in his own right. "*We owe him a huge debt*", he acknowledged.

This experience with Pierre Aigrain directed him firmly towards solid-state physics, instead of particle physics, the preferred field for many theoreticians. "*In particle physics, I had the impression that, although it raised many interesting questions, not much was actually understood, and that in the end no one really knew how to describe the interactions, except by adjusting parameters on an* ad hoc *basis. At the time, particle physics had a rather shaky aspect that I didn't like, unlike solid-state physics, which seemed to me to be on firmer ground. And also, Pierre Aigrain was a warm person, whereas the head of particle physics was intelligent, but cold. That was another factor*", he admitted.

So it was a year that considerably influenced Pierre-Gilles, although he did not devote all his time to the lab, unlike other interns, like Philippe Nozières, who worked every hour of the day and night, including weekends. Pierre-Gilles did not commit himself to that extent, because an attractive girl called Annie had just entered his life, and would remain part of it forever, a permanent and essential foundation.

Meeting Annie

Until this point, he had never been seen in a stable relationship. The day he answered a party invitation by saying: *"Yes, I'll be bringing Annie"*, everyone knew that it was serious. And in fact, from that day on, he was never without her. Pierre-Gilles was immediately captivated by this radiant and funny young woman, whom he had met by chance. Relaxing in a boat on the Bois de Boulogne lake with his two classmates, Favard and Guinnebault, they heard a voice calling: *"Pierre! Pierre!"* A girl was waving in their direction. It was Annie, who was walking around the lake with two female friends and had recognised Pierre Favard, one of her brother's pals. The young men rowed to the bank and invited the girls on board. It was an afternoon of laughter and fun. *"I was attracted to him immediately. He was gorgeous. But I should have known better, because he immediately made me row…"*, Annie smiles. She was under his spell but, she thought, had kept it well hidden. In fact, everyone could tell that something was going on between the two of them. The afternoon came to an end, and they had to part. But Pierre-Gilles and Annie would soon see each other again, with increasing regularity.

Their first dates took place under close surveillance. Anne-Marie Rouet — Annie — was a girl of good family, from a *"middle-class, provincial Catholic family"*. *"We were petit bourgeois, but in spring I had to put on my little tailored dress and a little hat to go to mass"*, she recalls. *"I was educated by nuns, like all the girls from my background, a couple of years here, four years there, wherever my father's work took him."* He was a magistrate, and the family had moved often, to Saint-Amand, to Clamecy, to Montargis, etc., before settling in Neuilly a year earlier. Since Annie wasn't yet 18, it would be inappropriate for her to go out alone with a boy, especially as Pierre Favard had warned her brother: *"Don't leave your sister with lanky de Gennes, he's not too be trusted."* So big brother acted as chaperone. However, Annie invented ways to escape his surveillance. She had always been self-willed. At the age of eight, she refused to wear lace collars (*"It wasn't my thing"*); at 15, she no longer wanted to go on holiday with her parents; so at 17, it was out of the question to have her brother hanging around! She found clandestine ways to meet her boyfriend. *"Pierre-Gilles used to come and fetch me on his motorbike when I was working for a stockbroker in Rue des Mathurins. But there were street girls nearby who found him very good-looking on his bike, so I would hurry down to make sure he didn't get sidetracked",* she jokes. She would like to have studied at the Beaux-Arts, but *"it wasn't something you did in my family",* so she was studying for a law degree while *"doing temporary jobs to be independent".* Later, she would work for a few months at the Père Castor school, an experimental school founded by the creator of the Père Castor publishing house, located on Boulevard Saint-Michel. The lovebirds went out together a lot, to the *Écluse,* a cabaret by the Seine, to *Studio Bertrand* or the *Feuillantines,* movie houses that showed art and experimental films. They also enjoyed walking in the Forest of Fontainebleau.

All this time, Annie's parents thought that she was with a girl-friend, until the day when Pierre-Gilles came to fetch her at Neuilly, to take her to a picnic in Marly. The time had come for introductions. Her mother's eyes widened when she saw this youngster. *"Not only did he have long hair, but he was wearing very short shorts",* recalls Annie. This was too much for her mother, who laid down the law that evening: *"That's the last time he sets foot in this house."* But Pierre-Gilles gradually won her round and become part of the family. Indeed, he was not only captivated by Annie, a whirlwind of high spirits, but also by family life in the Rouet household. *"They were an extremely pleasant family. They got on very well and never argued",* he recalled. The house was cheerful and welcoming. Every Sunday lunchtime, Annie's mother would feed tablefuls of cousins and friends. After the meal, Annie's father would play the piano. *"My father was a warm and generous man, who loved telling us funny stories. I had a very happy childhood, always with people around, unlike Pierre-Gilles",* she recounts. She had a generous nature, which came from her father, and believes that her stability and enjoyment of life came from her happy childhood. *"Annie is someone it's hard not to like!",* asserts a family friend.

A year later, a happy event was expected, which somewhat curtailed the engagement. Annie's mother came round to the idea of marriage, unlike Yvonne de Gennes, who would not hear of it. Out of the question that this daughter of Catholic *petits bourgeois,* whose father was "only" a magistrate, should wed her son! But could any girl have found favour in her eyes? Almost certainly not. In any case, she did everything to separate the couple, even employing a beautiful Swedish au pair girl to seduce Pierre-Gilles and writing very harsh letters to Annie, making her life very difficult. Pierre-Gilles waited for things to calm down. *"Initially, relations between my mother and Annie were very tense"*, he acknowledged, *"but they subsequently improved, thanks to Annie, who agreed to take her into our house when my mother reached an age when she could no longer live alone. It was a fine thing to do, because my mother was not an easy person."* Fortunately, Pierre-Gilles's aunt Edmée became fond of Annie. *"Cherish her! You won't find another like her"*, she enjoined her nephew. She took it on herself to organise the engagement party during the Easter vacation, since Yvonne de Gennes refused to have anything to do with the marriage. Annie was presented to Pierre-Gilles's extended family. One cousin enquired about her Hugenot descent: *"What exactly is your future wife's background?"*, he asked the husband-to-be, who took great pleasure in replying, with a wide smile: *"She's from Berry."* All his École normale classmates were there. *"Pierre-Gilles was the first of us to get engaged and to marry. It didn't surprise us that much: he never did anything like everyone else! And that's why we liked him so much…"*, recounts Roland Omnès.

The wedding took place on June 2, 1954 at the Temple de Neuilly. It was an unostentatious event, attended just by close family and a few friends. Although a Catholic, Annie had agreed to get married in the temple to please her mother-in-law. After the ceremony, the guests came together for a simple lunch at Annie's parents' house. Then the newlyweds set off to Provence for their honeymoon. *"It was the first time that we had been together all the time. We had never lived together"*, she remarks. And they got on famously. They stayed in a picturesque inn and visited Tourrettes-sur-Loup, Saint-Paul de Vence, etc. Back in Paris, they moved into an attic room above Annie's parents' apartment in Neuilly. Unfortunately, she learned that she would have to spend her time lying down to prevent a premature birth. *"He kindly stayed with me and worked at home, so that I wouldn't be on my own"*, she recalls. He was writing his degree dissertation on "transient states of carrier diffusion in a semiconductor film", using Aigrain's thesis to help him with the maths. Annie was careful not to disturb him. She instinctively understood that she should not restrict his freedom and a *modus vivendi* was established between them, like a tacit agreement, which would contribute to the strength of their bond. So when, having got his degree, he prepared to go on vacation leaving her bedbound, she did not complain.

On the Roads of Yugoslavia

Pierre-Gilles chose an incongruous destination for his vacation: Yugoslavia, where, under Tito's regime, tourism was almost non-existent. *"The country was almost unknown at the time, and surfaced roads were rare. But we had a fantastic trip."* He travelled with Jacques Signoret and Roland Omnès, in a Citroen 2CV borrowed from Jacques Signoret's aunt. In Zurich, they took a wrong turning. Too bad, they would visit the Italian Lakes instead of travelling through Austria! They travelled across northern Slovenia, then on to Dubrovnik, Kotor, Mostar, etc. They passed through villages where the children had never seen a car before. They mostly drove along rough tracks, with the result that one morning a stone ruptured the car's petrol tank. They began by trying to plug the hole with paper and shaving foam, but without success. Pierre-Gilles took charge: he emptied an oil can, filled it with petrol, strapped it to the front fender and ran the fuel pipe into it. This improvisation, reminiscent of Nicolas Bouvier at the wheel of his Fiat Topolino in 1953, carried them as far as a garage where they could get the vehicle fixed. Pierre-Gilles was not always quite as inept as he liked to pretend! The rest of the trip went smoothly.

At the end of each day, they would find a quiet spot to pitch their tent, near a lake or river, and would bathe and fish. *"We caught crayfish with forked sticks I made. It was a fairly primitive technique, but it worked well: we gorged ourselves on crayfish"*, he recalled. One day on a boat trip they heard a loud splash, looked over the side and were amazed to see a pike on the surface, biting its tail. *"The naturalist Jacques Signoret started to explain that it must be the pike mating season, but Pierre-Gilles wasted no time, manoeuvred the boat alongside the stranded pike, reached down and hauled it into the boat, and finished it off with a knife"*, recounts Roland Omnès. *"We ate it the same evening, discussing the symbolism of the snake Ouroboros in the hermetic tradition, which led us on to cyclical structures in chemistry* (Author's note: the serpent Ouroboros biting its tail is purported to have inspired the chemist August Kekulé, discoverer of the cyclical structure of benzene). *"* In this way, they spent their evenings chatting around the fire, Pierre-Gilles always hanging on the words of Roland Omnès: *"He might talk about the god Mithras incarnated in a bull, or equally well of Jung and psychoanalysis."* On some evenings, Jacques Signoret would sing mediaeval troubadour songs to entertain his companions. They attended village festivals to watch the Bosnian girls dancing in their traditional costumes, with pheasant feathers in their red velvet hats. But it was already time to return to France. The young physicist began his final year at École Normale, where the focus was now on preparing for the *agrégation*, a competitive civil service examination in France's education system.

Preparing for the *Agrégation*

"Agrégation year was a good opportunity to absorb all the basic physics courses", he commented. Indeed, apart from mounting experiments in the lab, he concentrated on revising for the exam. *"We spent hours trying to synchronise the flashes of a stroboscope with individual drips from a faucet, in order to be able to observe a "frozen" drop. It sounds simple, but it was hard, since the drops didn't fall regularly, perhaps because of the vibrations in the lab"*, he complained. When not in the lab, he immersed himself in *"basic physics"* in the textbooks of Landau and Lifchitz, which he read in English. These were not really "basic" texts. Lev Landau, an eminent Russian theorist, would win the Nobel prize in physics in 1962 for his work on the superfluidity of liquid helium. He headed a Russian school, and in 1949 had begun this imposing treatise on theoretical physics, a masterly work of a very high level, with one of his students, Yevgeni Lifchitz. *"It contains a cleverly distilled combination of calculation methods and physics, of extraordinary inspiration. For example, for a problem of particles in a potential, where many authors enter into long formal developments, Landau has a simple and brilliant*

argument that takes up a few lines. That's where you can see the mark of a great man. This physics treatise is six or seven volumes long. After his death, his students wanted to add a few volumes. Yesterday, I looked at one to check a particular point, but I found massive calculations: Landau's spirit was no longer there. I didn't find the arguments I needed, so I had to reconstitute them myself", he reported. Pierre-Gilles would return to this monumental work throughout his life, reading and rereading it to *"clarify his ideas"* and urging his students to do the same.

He looked up from his books at the end of 1954, when Annie came to the end of the fateful nine months. An appointment had been made at the very smart Marignan clinic, in the 8th arrondissement, on Yvonne de Gennes' recommendation. He drove her there, then went to wait at the... Palais de la Découverte, a stone's throw away. His son, Christian de Gennes, was born in December 1954. Pierre-Gilles and Annie moved into an apartment on Rue du Mail, near Place des Victoires. Annie had stopped working and took care of everything, in particular arranging an office for her hard grafting husband in a small, bright corner room. This suited him perfectly in the run-up to the *agrégation*. From time to time, he would go out, a textbook in his hand. *"I can still see myself pushing Christian down Boulevard Saint-Germain in a buggy before the exam"*, he remembered.

The results of the competition brought no surprises. Philippe Nozières and Pierre-Gilles were a hair's breath apart, with respectively 403 and 402 points out of 500, behind Jean-Pierre Barrat, first with 450 points (the fourth had 350 points). *"Jean-Pierre Barrat was a 'Phoenix of learning'"*, recalled Pierre-Gilles de Gennes. Philippe Nozières explains: *"He was the cleverest. He got into the École Normale at the age of 18, after only one year of classe préparatoire. Like Pierre-Gilles, he had a touch of genius. He then produced a brilliant thesis on optical coherence in Alfred Kastler's lab, supervised by Jean Brossel: his work won him the Optical Society of America Prize — exceptional for such a young researcher — and the School of Claude Cohen-Tannoudji, Nobel prize for physics in 1997, is his legacy. He then worked at the University of Caen."* In any case, this ranking was of only anecdotal interest, since they had already chosen their next step after school, having no shortage of options. *"The future was very open. There were places everywhere…"*, acknowledges Philippe Nozières.

Pierre Aigrain had urged Philippe Nozières and Pierre-Gilles to do a doctorate in the US. But only Philippe Nozières took the bait, setting off for America *"as excited as a gold seeker"* to develop a theory on the properties of free electron gases at low temperatures in David Pines' laboratory in Princeton. Pierre-Gilles was briefly tempted to work on band structures with Walter Kohn, a physicist already famous in semiconductors (he would receive the Nobel Prize for chemistry in 1998), but fearing that the work might become repetitive, he eventually turned down the offer. Some thought that he refused the position because, as a young father, he was no longer free to cross the pond. Subsequent events would prove them wrong.

Pierre-Gilles did the tour of the cutting-edge research labs to identify promising fields under top lab directors. He was hesitating between three subjects: the electronic structure of metals with Jacques Friedel, at the École des mines, nuclear magnetic resonance with Anatole Abragam, at CEA Saclay, or magnetic structure with neutron scattering with André Herpin, also at CEA Saclay. Actually, this institution, built in 1952, had considerable resources and included brilliant young researchers like Jules Horowitz, Anatole Abragam, Albert Messiah and Claude Bloch, most of them recent initiates into the "new physics" in the US. The young graduate saw their presence as evidence of dynamism and a guarantee that he would encounter new ideas. *"I phoned Jacques Yvon* (Author's note: Director of the Department of Theoretical Physics, who would be appointed AEC High Commissioner in 1970) *to tell him I was thinking of joining, and he replied: "I'm sending you a car with a chauffeur." My bluff was called"*, he remembered. The tour of the labs, the reactor and the accelerator also impressed him: *"It felt that they were building the future of research."* His decision was taken: he would do his PhD in the theoretical physics department headed by André Herpin, at CEA Saclay. *"I made up my mind on a superficial impression, but it proved a good decision."* Firstly, he would get on well with André Herpin, an open-minded and curious man. Secondly, *"as luck would have it, I saw a lot of Jacques Friedel and Anatole Abragam over the next few years, people I was planning to work with."* The die was cast.

Chapter 5

Research Apprentice

In October 1955, Pierre-Gilles de Gennes began his career in research. The years that followed were years of "firsts": first scientific article, first international conference, etc. The young man was joining France's prestigious "Atomic City", CEA Saclay, in the Paris suburbs, home to one of the country's two neutron reactors, EL2. At the time, there were only about a dozen neutron reactors in the world, the first having been built in 1943 at Oak Ridge in Tennessee, to make the plutonium needed for the first American atomic bombs. After 1946, reactors were also used for more peaceful purposes, to produce neutrons for the study of matter. Because neutrons are electrically neutral, they can penetrate matter without being blocked by the Coulomb barrier. This means that when fired at a sample, they collide with the atoms in it and scatter, i.e. are diverted from their trajectory, without being captured as they leave the sample. They then form a "scattering spectrum" (a diagram representing their direction and energy), which physicists can use to map the arrangement and motions of the atoms in the sample. Neutron scattering is therefore a precious resource for exploring matter.

"For its time, it was a very powerful tool, which attracted interest from many physicists", commented Pierre-Gilles de Gennes. *"But I*

didn't go into it, because I thought it was already a closed field. There were experiments to be done, but from a theoretical point of view there wasn't a lot of work left. By contrast, Bernard Jacrot was starting to do scattering experiments in magnetic media..." Bernard Jacrot was a thirtyish scientist in France's elite École Polytechnique, who belonged to André Herpin's group — *"an essentially experimental group, although he himself was a theorist"*, according to Pierre-Gilles de Gennes — who, with his colleague Magda Galula, was running France's very first experiments in neutron scattering from magnetic materials on the brand-new EL2 reactor. Although neutral, the neutron actually has a "magnetic moment" through which it can interact with the magnetic moments of atoms, which can be represented as small magnets carried by the atoms (for the sake of simplicity, they will be referred to as "spins" in the rest of this chapter) (A5-1). This means that the neutrons make it possible to "see" not only the atoms, but also their spins. *"Our experiments seemed to interest him. So Pierre-Gilles de Gennes started to work on the theory"*, recounts Bernard Jacrot. Simply speaking, the young researcher was given the job of determining how to deduce the arrangements and motions of the spins from the shape of the scattering spectrum.

For his first foray into research, therefore, he chose a subject that was both pioneering and promising, in two respects. Firstly, neutron scattering was a cutting-edge research technique. Secondly, the subject was right at the heart of the physics of its time: the discipline had undergone a revolution in the early 20th century with the discovery of the laws of statistical physics and quantum physics. Now, the task was to apply them at microscopic level — to atoms and electrons — in order to explain the properties of a material at macroscopic level, notably its electrical and magnetic properties. In fact, the study of the arrangement of spins as a key to the magnetic behaviour of materials was one of the most fashionable topics in solid-state physics at the time.

"Ideal Theorist"

Pierre-Gilles was aware that he was in the engine room of contemporary science, and was as excited to begin his research as on his first visit, when Jacques Yvon, director of the department, had sent a chauffeur-driven car to fetch him. Bernard Jacrot, *"very kind to the callow youth I was at the time"*, showed him the imposing neutron reactor in its immense, ten metre square concrete cube, containing three tonnes of uranium and seven tonnes of heavy water. The young man found out how it worked and demonstrated that, novice though he was, he knew a thing or two about neutrons. Not surprisingly, as he had spent a good part of the summer immersed in books. *"As soon as I knew that I would be joining CEA Saclay, I began to think about neutron scattering. No one had asked me to do it. I was just curious. I worked over the summer, but I did that most every summer after the baccalaureate, and later on too."* So in the summer of 1955, instead of climbing mountains or hunting marmotte, as he usually did in Orcières, he delved into the work of Léon van Hove, a Belgian physicist who had written the seminal article establishing the general formalism of neutron scattering. *"He was the great prophet of that whole subject and his article is still the number one reference in the field"*, explained Pierre-Gilles de Gennes. The young man took a good look at the reactor, but would subsequently have little to do with it, spending most of the time at his desk. Nonetheless, he did not stay shut away with his calculations. *"Bernard Jacrot and I saw each other all the time"*, he recalled. *"Our offices were a few yards apart. We talked very often. You could say that we formed a real theorist-experimentalist duo"*, agreed Bernard Jacrot.

For the moment, he was just a greenhorn, with everything to prove. He knew that his career began here. Since Bernard Jacrot and Magda Galula had began their first experiments on iron, he applied Léon van Hove's model to this "ferromagnetic" material, focusing in particular on fluctuations in the spins caused by temperature. The spins are not fixed. At low temperature, they are all aligned within contiguous domains, called Weiss domains, which subdivide the sample like pieces of a puzzle. Since the direction of the spins varies slightly from one domain to another, it is only when a sufficiently strong magnetic field is applied that the spins of all the domains line up in the same direction: the iron then becomes magnetised. However, if the temperature is increased, the spins gradually lose direction, like swinging compass needles, causing strong fluctuations in local magnetisation. Above the so-called Curie temperature, the spins become completely disordered: the iron loses its magnetism, moving into the "paramagnetic" phase. Since low-temperature neutron scattering by iron had already been extensively described, Pierre-Gilles studied the case of high temperatures and the area — called the critical region — around the Curie temperature.

He began by trying to form a picture of the disorder that pertained at high temperatures. The situation was complex: each spin was influenced by the intense magnetic field created by its neighbours but, as the little magnets constantly changed direction, the field they produced also varied constantly! To model such a situation was a rather like trying to grab hold of a piece of soap in the bath. Pierre-Gilles overcame this apparently insoluble problem by using a mathematical method called the method of moments and managed to calculate the form of neutron scattering spectra at high temperatures. However, he hit a problem: his result was in contradiction with one of the leading lights of physics, the American John Hasbrouck van Vleck (who would be awarded the Nobel Prize in 1977). *"He was one of the first in the US to understand quantum physics, and immediately succeeded in applying it to numerous situations"*, observed Pierre-Gilles de Gennes. In fact, the difference in the results arose from the fact that, unlike van Vleck, the young physicist had taken into account the interactions with several neighbours, not just one. If the American's hypothesis was right, Pierre-Gilles thought, the dynamics of the spins in relation to each other should remain the same in all circumstances, and therefore be independent of the wave vector of the neutrons probing the magnetic properties of the sample, which he did not believe.[1] Could he, a young postgraduate, have put his finger on a mistake by this giant of physics? It seemed unlikely... He redid his calculations, with the

---

[1] The wave vector of the neutrons represents the direction and wavelength of the incident neutrons.

same results. He became convinced that van Vleck had made an error in neglecting long-distance interactions between spins, which clearly had to be included in the calculations.

To set his mind at rest, he asked Daniel Cribier, an experimentalist who had joined Bernard Jacrot, to confirm that the spin dynamics depend on the neutron wave vector, as he thought. They decided to conduct the experiment in manganese fluoride, a compound with a large spin, where the effect should be obvious. *"He asked me to verify that the spin dynamic was different with a small, medium and large wave vector"*, recalls Daniel Cribier. *"But every time, I got the same result."* The experiments confirmed van Vleck's conclusion! Pierre-Gilles couldn't believe it. Had he made a mistake? He plunged back into his calculations, but it was the experimentalist who discovered an artefact in the experiments: the samples contained protons which had distorted the results. Cribier redid the measurements, working at night to get a more stable electric current and, this time, the results proved de Gennes to be right, relieving him of further recalculations.

In 1958, the young man met van Vleck, who would acknowledge the value of his work. *"He was a pleasant, quaint old gentleman, with some surprising obsessions. For example, he knew all the train timetables by heart: he knew that in order to travel from London to Birmingham, you have to take the 18:57. He also loved American football. When we met again in Harvard, a few years later, he took me to a match — there were even cheerleaders — and explained the rules. By then, he was over 70, but still completely involved in the match. I was always fond of him; people called him Van. He had taught Philip Anderson, the leading figure in solid-state physics in my generation"*, he recalled warmly.

A Taste of Success

Describing disordered high-temperature spins, thereby correcting an omission by van Vleck, was a significant achievement. Pierre-Gilles recalled: *"André Herpin was following these high-temperature calculations with interest. We used to talk every week, or even every day in 1956, when we found ourselves stuck in Paris without fuel, following the*

Suez Crisis. We used to take the bus to Saclay from Denfert-Rochereau. We would often sit together, and talk about all kinds of stuff, my work, painting..." Bolstered by this first success, he began work on the description of the critical region, drawing on the work of Léon van Hove which predicted that a ferromagnetic element at the Curie point (a phase in which all spins are parallel) should start to behave similarly to a fluid at the "critical point", a particular stage (374.15°C and 218.3 atm for water) where a fluid is neither a liquid nor a gas (A5-2). This strange "fog" is the source of large fluctuations in density and therefore in refractive index. It means that under illumination the light is scattered in all directions and the fluid becomes milky in appearance (opalescence). According to Léon van Hove, a similar phenomenon should occur at the Curie point in a ferromagnetic material when it becomes paramagnetic, the density fluctuations being replaced by magnetic fluctuations, and the light by neutrons.

Pierre-Gilles formulated an approximation and demonstrated, step-by-step, how and why opalescence occurs in ferromagnetic materials, as revealed by neutron scattering. He extended this result to antiferromagnetic materials, substances (like chrome or platinum) in which the spins are aligned similarly to ferromagnetic materials, but nose to tail, i.e. facing alternately in one direction and the other (as long as the temperature remains below the "Néel temperature", equivalent to the Curie point). This was a significant result, but he still wanted to understand how the spins change as the Curie point approaches, in other words how the phase transition between order and disorder occurs. His method was simple. He looked at a spin at a given moment, when its orientation is set in one direction (the other spins have no set orientation because of the disorder that arises at high temperature). Once the spin became "free", the young physicist focused on the directions of the other spins. On average, after a certain time, they tended to equalise (since the spins can go in any direction at high temperature). Comparing this process to a standard diffusion phenomenon, for example the spread of a perfume spray through a room (after a few minutes, the attenuated smell of the perfume will be the same throughout the room, since the concentrations tend to equal out), he simplified the problem to... a simple calculation of the diffusion coefficient (and showed that the coefficient is cancelled out at the Curie point). Establishing a description of the spin dynamic in the critical region was not the achievement of a novice, but reflected a precocious and acute sense of analogy. It brought him to the attention of the specialists in statistical physics at Harvard.

However, Pierre-Gilles was unable to calculate the diffusion coefficient as he wanted, because he was not satisfied with the models he was using. So he build a new one, inspired by... Richard Feynman (A5-3). *"It was an amusing transposition of one of Feynman's ideas to a different context. I had come across this calculation in articles of his I read at the École Normale. It shows how much he influenced me as a physicist."* With the new model, he could re-demonstrate the existence of "spin waves", collective excitations caused, broadly speaking, by the interactions between spins. It was almost as if they were linked by little springs — when one switched it took the others with it — which formed a sort of sea in which waves could appear. Case closed: it was proof that his model held together perfectly.

Tributes from his Elders

In just a brief time, Pierre-Gilles had achieved some substantial successes. *"At the time, it all looked relatively new. But now, it just strikes me as tinkering around the edges"*, he smiled. So he wrote his thesis, which proved rich in analogies and often emphasised the physical mechanisms involved rather than the formalism: *"(It) will be enough, for our purposes here, to employ the simplest statistical processes (...) and only to make a fairly brief study of the orders of magnitude: our intention is above all to give a physical idea of the kinetics of the fluctuations"*, he writes for example in chapter 4. Already in this manuscript, we can see the early outlines of a style that would become characteristic. He presented his thesis on December 9, 1957 at the Sorbonne, after just two years of research (he then remained at CEA Saclay with the status of research engineer). The members of the assessing panel were all great names of French physics: Francis Perrin, Professor at the Collège de France and AEC High Commissioner, Yves Rocard, Jacques Friedel and Louis Néel. The presentation process was a formality, since in everyone's opinion the young physicist's thesis was "*a valuable synthesis of the state-of-the-art in magnetic neutron scattering*", as Jacques Friedel would describe it. From the beginning of his thesis, however, Pierre-Gilles felt that he was doing no more than following the trail blazed by Léon van Hove, developing calculations that any theorist could do. He had no intention of being a backroom boy all his life, content to explain the ideas of others. His aim was to formulate his own ideas. From 1956, therefore, before even finishing his thesis, he began to look for topics where he could show what he was capable of, or "*shine more*", as some physicists said.

Every week he browsed carefully through the scientific journals and was happy to talk to anyone in his quest for new subjects. He listened carefully as a young experimenter, Maurice Goldman, described his experiments running a strong electrical current through a bath of melted gallium to separate its isotopes. On his own initiative, driven by curiosity, Pierre-Gilles tried to elucidate the phenomenon and construct a model (A5-4). *"For the first time, I felt I had solved a problem on my own"*, he recalled with satisfaction.

The young physicist was also ready to talk to more experienced theoreticians, for example Claude Bloch, a specialist in nuclear physics. *"He was an imaginative and talented physicist, who unfortunately died very young, without leaving the body of work that might have been expected of him"*, mourned Pierre-Gilles de Gennes. *"I remember asking him a question on the collective motions of nucleons (protons and neutrons) in the nuclei, which impressed him. I had discussions with him that went beyond my own limited little field."* Nothing pleased the young theorist more than recognition by his elders. *"I also knew Jules Horowitz, the big-name in French nuclear science. He was an incredible worker: he used to eat out of a mess tin on his desk so as not to waste time."* The theorist would gradually gain their respect, emerging as one of the rising stars of the younger generation. He had become a physicist worth watching.

"I made some really valuable contacts at the CEA, the most valuable of all with Abragam, an incredible source of inspiration", he noted. Anatole Abragam was the researcher who had constantly interrupted during his paper at the École Normale Supérieure two years earlier — a "terror" for many young physicists! *"When he came to a seminar and failed to find it interesting, he would leave after ten minutes. He didn't mean any harm by it, but his departure disconcerted many a speaker"*, laughed Pierre-Gilles de Gennes. Anatole Abragam would often burst into his office to ask him questions, to test him or for distraction... *"One day, he came to ask me a question about a problem of nuclear relaxation due to a paramagnetic impurity. He was carrying an article by a Georgian physicist, and asked me to repeat the demonstration."* However, the article proved incomprehensible. The young theoretician had absolutely no idea what the authors were trying to get at, but persevered, aware that Abragam was testing him. The subject suddenly reminded him of a more general situation described by Enrico Fermi, the famous American physicist. He looked out the article and identified a link. A few weeks later, he had elucidated the mechanisms, simplified the demonstration and even extended the description to the more realistic case of multiple impurities. This success won him the respect of the demanding Anatole Abragam, as well as a privilege: correcting the massive tome his mentor was in the process of writing, *The Principles of Nuclear Magnetism.* *"It's very clear, impeccable"*, he promised in chorus with the other two proofreaders. A week later, Anatole Abragam was back. *"This chapter is completely wrong and you didn't notice!"*, he exclaimed. Indeed, one of the initial hypotheses was wrong. Pierre-Gilles was mortified, but Anatole Abragam did not lose confidence in him.

Missed Discovery

One day, Anatole Abragam introduced him to Jacques Friedel, a consultant at the CEA, whom the young man knew only by reputation, as one of the leading lights of solid-state physics in France, the author of numerous articles on the electronic structure of alloys. After this, Jacques Friedel would frequently come by for a chat.

"He was interested in all sorts of things. He was familiar with everything I had done. Then, I started to work on the metals called rare earths[2] which have "delocalised" electrons scattered by the magnetic moment of atoms, in a similar way to neutrons scattered by the magnetic moments of iron, which Pierre-Gilles de Gennes had worked on for his thesis. Since the research had not been done on rare earths, I suggested he should do it...", recalls Jacques Friedel. *"A fortnight later, he came to see me at the École des mines (where I was working at the time) and gave me the written article (with my name on it, which was generous, since I had only suggested the idea). The problem wasn't that hard, but it still required thought and a degree of inventiveness. He did the whole thing in two weeks."* Pierre-Gilles wrote another two articles on magnetism in the rare earths (A5-5). *"By studying the interactions between rare earth atoms through an approach recently developed by Kittel and his collaborators, I noted that there were numerous fundamental states that were very close to each other, in other words that the system could "decide" to be ferromagnetic or antiferromagnetic, or else to adopt bizarre forms of antiferromagnetism. I mentioned it, but without going any further..."*, he confessed. However, it was more than a minor detail and would result in a significant discovery, one that would be made by... his student.

Indeed, André Herpin had given Pierre-Gilles his first student, Jacques Villain. *"To begin with, I suggested that he should do calculations of critical point scattering on more complicated systems than the simple antiferromagnetic materials I had studied in my thesis"*, recalled Pierre-Gilles de Gennes. Jacques Villain bent valiantly to the task, obtaining equations that led him to think that maximum scattering could be attained in a different spin configuration than in ordinary antiferromagnetism... What if the spins were tilted? *"I went to see Pierre-Gilles de Gennes in his office to tell him that I thought that we don't necessarily get what we expected and explained my idea. After 20 seconds of reflection, he said: "Yes, I think that's an excellent idea."* In fact, in certain cases, instead of arranging themselves into two antiparallel sublattices as they do in an antiferromagnetic material, the moments can form a spiral: the

---

[2] These are the elements with atomic numbers between 57 and 71.

magnetism is helicoidal", recalls Jacques Villain (A5-6). Pierre-Gilles categorically refused to cosign the article on this discovery: *"All the merit belongs to Jacques Villain"*, he insisted. But the adventure did not end there. *"As it happened, experimentalists in our lab were trying unsuccessfully to determine the magnetic structure of a manganese-gold alloy. Pierre-Gilles de Gennes immediately made the connection: "Villain has had a weird idea that might explain the structure", he suggested. And in fact, it worked perfectly and, for the first time, we saw rotating magnetic moments!"*, recalls Jacques Villain with pleasure.

Several Irons in the Fire

For Pierre-Gilles, operating at cruising speed meant having several irons in the fire. His career would be characterised throughout by this need to work on different fronts. Between 1956 and 1959, he continued to diversify his research topics, as if afraid of becoming trapped in a single field. He kept his antennae tuned for potential topics of research through discussion or reading. In 1959, he started thinking about a situation where A and B atoms are randomly arranged on a lattice, with A being active (e.g. a conductor) and B inactive (e.g. an insulator).[3] When few in number, the A atoms form clumps that are isolated from each other: the sample is an insulator. When there are enough of them, they form a sufficiently large mass to span the sample, which then becomes a conductor. Pierre-Gilles showed that there is a critical concentration, arising solely from the geometry of the system, which induces a phase transition: the inactive lattice becomes active. *"For the first time, I had the sense of discovering something big and significant."* He sent this article to Philip Anderson, the great American theorist of solid-state physics, who had already studied random lattices. Anderson immediately sent back his own articles on the subject, which Pierre-Gilles hastened to read. A footnote drew his attention to the work of an English

---

[3] The idea came to him following a discussion with Jacques Friedel, who was wondering in what conditions an alloy made of spin-carrying (ferromagnetically paired) A atoms and spinless B atoms would make a Curie point possible.

mathematician called John Michael Hammersley, who had unfortunately developed similar calculations in seeking to find out when a
gas mask would become clogged through accidental dust inhalation.
"I realised that I wasn't the first to get there...", he reported. The young
theorist was disappointed. Without realising it, he had reinvented
the phenomenon of percolation. However, since his approach took
account of the system's geometry and real physical properties, such
as conduction or magnetism, he still published his calculations.
"This was the first work on percolation that was not purely speculative",
confirms Jacques Friedel.

This original publication by Pierre-Gilles sunk without trace,
until it was unearthed and its originality acknowledged in the late 70s.
At this time, percolation became a primary tool in statistical physics
for the study of situations as diverse as the formation of a gel, the
flow of liquid in a porous medium or the spread of fire in a forest,
but also in social sciences and economics, for example to describe
the way information travels through a network. Pierre-Gilles, who
often used it himself, was one of the contributors to the success of
this universal tool.

Whatever he did, his boss André Herpin gave him free rein.
*"I felt encouraged in this diversity, but all my topics remained close to the con-
cerns of the institution: magnetism and solid-state physics, in general, and
neutron scattering"*, he insisted. André Herpin never intervened in his
decisions, except once, when his suggestion led to disaster. That day,
as he occasionally did, Herpin joined Pierre-Gilles, Bernard Jacrot
and Daniel Cribier for lunch. They liked to go to the small restau-
rants around Saclay, the *Hôtel de la poste* or the *Auberge de la croix de
Bures*, where Pierre-Gilles never missed the days when they served *terrine*. *"It was a time when the Chevreuse Valley was not overrun. At the
restaurant, they used to bring us an immense home-made* terrine *and let us
eat as much as we wanted! In those days, I had a colossal appetite and could
eat a horse. Sadly, there came a point when they started serving us the* terrine *in individual portions on small plates, to our great disappointment"*,
mourned Pierre-Gilles de Gennes. Foreign guests often ate with
them. On this occasion, one of them, Roman Smoluchowski, a met-
allurgist from Princeton, complained that there were no

experimentalists available to study the damage caused to a tungsten wire by alpha radiation emitted by radioactive nuclei.[4] André Herpin turned to Pierre-Gilles and suggested: *"Why don't you give it a try?" "Sure, why not?"*, he replied. It would prove a more difficult task than he imagined. *"The whole thing turned into a nightmare because, since I wasn't part of any experimental lab, I didn't have even the simplest tools. I had to beg and borrow them where I could."* When the experiment was finally ready, he lined up the tungsten wire with the beam of the Van de Graaff accelerator, but the brilliant theorist was a clumsy experimentalist, and put one eye directly in the beam. He left the lab in an ambulance. It was serious — there was a risk that he might lose an eye. He underwent several medical examinations, with positive results. Back in the lab, he resumed his experiment, with more care, but with no success. *"The wire became cherry red, which meant that most of the damage that appeared in the wire took the form of heat: in fact, the experiment was meaningless. We hadn't thought it through"*, he recalled, regretting all the wasted days spent preparing this pointless experiment. *"The conclusion of all this was that it was better not to encourage me to do experiments."*

Cheerful Evenings

His research left him little time for family life. He would get home late to the apartment on Rue du Mail, where Annie always greeted him with a smile despite her own very busy days: after Christian, born in December 1954, Dominique and Marie-Christine were born respectively in May 1956 and January 1958, and she took care of everything. *"Pierre-Gilles never fed a baby or changed a diaper. He wasn't interested, and one might say that he did have other things to do. It's simple: in day-to-day life, he never did anything"*, she laughs. She took charge of everything — from getting the car serviced to buying his clothes — and was perfectly content to do so. *"You don't get to the level he achieved*

---

[4] Alpha radiation, emitted by uranium or radium nuclei, consists of two protons and two neutrons. Being large and heavy, it does not easily penetrate matter, but it has strong ionising force.

by half measures", she asserts. In the evening, she still had enough energy left to entertain — she loved cooking — usually friends from École Normale days. *"We had stayed very close and we spent some great evenings at Pierre-Gilles' place. He had an extraordinary apartment, ornate and crooked, with an interesting structure, because you could roll a child's toy car from anywhere, and it would always end up in the same place"*, recalls Philippe Nozières. There were animated discussions about the existence of the mysterious minimum-energy point. They also revelled in the bizarre sword duel fought by the ultrarich Marquis de Cuevas — which made headline news in the spring of 1958 — in which Lucien de Gennes, Pierre-Gilles' physician uncle, had been chosen to record first blood.[5] *"We didn't belong to the same milieu. We didn't see him often, because he was very busy, and he also didn't much like my mother, whom he found too talkative"*, reported Pierre-Gilles de Gennes.

Everything was fine in the family, except that Yvonne de Gennes gave her daughter-in-law a hard time, inviting herself and her friends to tea and criticising the state of the cupboards. She would stick her nose into everything... Annie de Gennes recalls one example: *"When they were ill, I used to take the children to the nearest paediatrician, who was on Rue de Rivoli. She didn't approve and asked advice from Lucien de Gennes, who told her that he was the best in the business. Only then did she stop carping."* She shared her exasperation with her husband, who just shrugged: that was the last she said of it. Every Sunday, the family went for lunch to Annie's parents in Neuilly, where the atmosphere was always pleasant, but immediately after the meal Pierre-Gilles would leave them and return to his study on Rue du Mail to work until evening.

On the Lecture Circuit

The young theorist had enough results under his belt to present them to his peers at national and international conferences. He gave

---

[5] This duel between the dancer and choreographer Serge Lifar, and the Marquis de Cuevas, director of the Monte-Carlo Ballet, was stopped when first blood was drawn.

his very first lecture at the National Conference on Magnetism in Strasbourg in 1957. He drove there with Jacques Friedel, so absorbed in discussion that they scarcely noticed the length of the journey. At this time, Pierre-Gilles was starting to take an interest in superconductivity, the mysterious phenomenon that had kept all the world's physicists guessing since its discovery in 1911. When certain metals are cooled to a very low temperature, electric current flows through them without resistance: the samples become superconductors. Pierre-Gilles had no qualms about his lecture. Unlike many French researchers at the time, he was quite comfortable lecturing and answering questions in English. However, his talk about rare earths made little impression. A senior Dutch physicist, Cornelis Jacobus Gorter, even deemed his research completely pointless, going so far as to ask the Saclay physicists: *"Why on earth did you bring this guy to an international conference?"*

Pierre-Gilles made no more impact at the symposium on inelastic neutron scattering in solids and liquids, held in Vienna in October 1960 by the International Atomic Energy Agency. The atmosphere in the city sent shivers down his spine. *"The Russians had only recently left Austria. You could still see their mark in the streets, as well as traces of the war."* His paper on neutron scattering in liquids met with a critical reception. In it, he developed a formula that ignored quantum effects, stating that when a certain characteristic neutron wavelength becomes comparable with the distance between the particles in the liquid, the scattering spectrum of the neutrons carries a visible "trace" of them. An American physicist, Mark Nelkin, presented a similar study that included quantum effects. Pierre-Gilles was convinced that quantum effects are negligible in liquids, except in some cases such as helium and hydrogen. The two men went head-to-head. His opponent claimed: *"I would be amazed if anything at all came out of your description."* Experimental proof would not come until a few years later, vindicating... Pierre-Gilles. The "trace" that he had predicted is now called "de Gennes scaling". *"Mark Nelkin is a serious physicist who wanted to do the calculations properly, taking account of all possible effects. Sometimes this is necessary, sometimes it's not"*, he remarked. The French physicist always sought maximum possible simplicity, which could be summed up as "Why do more, when you can do less?", a position that would become his credo from the 1970s, with his use of scaling laws. This would lead some scientists to accuse him of a lack of rigour. However, all the skill of the theorist lies in separating the essential from the superfluous in describing a phenomenon. This requires common sense, intuition and a wide-ranging scientific culture. Pierre-Gilles excelled at this game to the point of becoming a master of the art of simplification.

Giants with Feet of Clay

It was in October 1957 that he took part in his first real international conference, on neutrons, in Stockholm. *"It was a small conference, with only around fifty participants. It was held the same week the Russians launched Sputnik and you could feel a degree of American panic to see the Russians take the lead in space."* Whilst the scientists could talk of nothing but the Russian satellite, he was worrying over an anomaly that he had come across in an article by two respected British theorists, Roger Elliott and Walter Marshall. The article was about critical scattering by magnetic media, a subject that he knew inside out. Between lectures, he took them aside and quietly explained his objection, but the two theoreticians waved it off. Sure of his position, Pierre-Gilles was indignant not to be taken seriously. Months later, when Walter Marshall was lecturing about some new research, de Gennes — spotting a flagrant sign error — would stand up and point it out in front of the whole assembly. He was only 24, but he was no longer intimidated by his elders, however eminent. It was only much later that Roger Elliott and Walter Marshall would acknowledge their mistake.

In June 1956, he had been invited to a special lecture in Cambridge, entitled *"Electron theory of transition metals"*, organised by the great British physicist Nevill Mott, who wanted to gather a small circle of specialists — only some thirty physicists had been invited, including Jacques Friedel, Magda Galula and Pierre-Gilles (aged only

24), the only French representatives — to discuss "in private" the incredible results reported by Richard Weiss. This experimental physicist claimed to have used x-rays to map the positions of electrons around the nucleus. *"It turned out that Weiss's experiments and Mott's ideas were completely wrong. So the whole thing has been forgotten. It gave me a good sense of the uncertain nature of what we do. Even a great physicist like Mott. There was no bad feeling, but Dick Weiss must have felt uncomfortable... Mott had made a classic mistake: developing a model to explain a failed experiment"*, observed Pierre-Gilles de Gennes. It was a lesson that stayed with the young scientist: even giants have feet of clay, which is why, much later, especially after the Nobel Prize, he would always emphasise that he was not an oracle.

Conflict with Louis Néel

At the end of 1958, he attended another international conference, on magnetism, in Grenoble, which might have changed the course of his life. The conference's big name was Louis Néel, a specialist on magnetism (awarded the Nobel Prize for physics in 1970), who was working in Grenoble. Pierre-Gilles was very interested in the lecture delivered by the physicist Harry Suhl from *Bell Labs*. *"He had discovered a form of interaction which I later explored, called the Suhl interaction, an interaction between nuclear spins which occurs as a result of the electron spins of the material"*, explained Pierre-Gilles de Gennes (A5-7).[6] While this lecture influenced the physicist's subsequent work, it could not be described as life changing. What might have been was a proposal made to him at the postconference reception. His work had attracted the attention of Louis Néel, who took him aside and asked: *"What would you say to a position as lecturer in Grenoble?"* Initially, an offer like this, coming from the grand master of magnetism in France, dazzled the young scientist, who would never have dreamed of a lecturer's position (equivalent to a professorship today). *"So I more or less said yes"*.

---

[6] Nuclear spin is the spin that results from the spins of the protons and neutrons that form the nucleus.

Back in Paris, however, he sought advice from colleagues and some, including Anatole Abragam, advised him against moving to Grenoble, where he would never experience such a rich scientific environment as in Paris, citing the names of brilliant researchers who had sunk into oblivion in the provinces. This dimmed the young man's enthusiasm. After a few weeks of reflection, he finally wrote to Louis Néel turning down his offer. There was another reason for his rejection. He had great admiration for the future Nobel Prize winner, but was irritated to see *"researchers in Grenoble almost prostrate before him and citing him at every turn: it was practically a personality cult"*. Louis Néel bore him a grudge for several years over his decision — *"which I can understand"*, acknowledged Pierre-Gilles de Gennes — and tried to put a spoke in his wheel on various occasions, for example delaying one of his promotions when he was President of the National Universities Commission, or blocking funding when the young theorist was setting up his own lab in 1961. *"There was a period when relations between us were a bit cool"*, regretted Pierre-Gilles de Gennes. However, the two men would mend fences a dozen years later. *"Dear Monsieur Néel, It is both a great honour and a great pleasure to receive this "sum"* (Author's note: a series of articles) *from you. (...) Thinking back over the last twenty years, I somewhat regret not having found more occasions to ask for your opinion"*, he wrote on December 7, 1978.

Off to California

By contrast, in 1959, a NATO school on solid-state physics in Paris undoubtedly did impact on the young theorist's life. Between two classes, Jacques Friedel suggested to Charles Kittel, the great American specialist on magnetism: *"Why don't you take on de Gennes as a post-doc? He has several strings to his bow: magnetism, rare earths, percolation. He has also learned quite a bit about nuclear resonance with Anatole Abragam."* He had no trouble persuading Charles Kittel, who, back in Berkeley, invited Pierre-Gilles to come. The young man was overjoyed. Berkeley was one of the world's most active research centres. He was also excited by the prospect of a stay in California. He recalled magnificent landscape photos that he had seen in *National Geographic* as a teenager. But what about his family? He didn't want to drag the children with him. Uncomfortably, he presented Annie with a fait accompli: *"I am going to Berkeley. You can either come with me or stay with the children."* After a few days' thought, she decided to follow her husband and made the necessary arrangements (Christian would stay with Yvonne de Gennes and the girls with their maternal grandmother). *"This attitude was typical of him: whenever he had a dilemma or an uncomfortable decision, he would cut straight to the chase"*, his wife recalled. Pierre-Gilles wrote to Charles Kittel giving his arrival date. At the same time, his future director was contacted by Berndt Matthias, a great superconductor specialist at *Bells Labs*: *"There's a young guy in France you should get over here. His name is de Gennes."* Charles Kittel was able to reply: *"It's in the pipeline, he'll be here in a month."*

Chapter 6

The Big League

The moment he arrived at Berkeley on January 1, 1959, Pierre-Gilles de Gennes felt at home. The campus overlooked San Francisco Bay. In the distance, the Pacific Ocean stretched out beyond Golden Gate Bridge. With their mixture of Victorian, Gothic and modern styles, the buildings gave the place a comical look — there was even a theatre copied from the amphitheatre at Epidaurus in Greece and a belltower identical to the *Campanile* in St Mark's Square in Venice! The French physicist's office was in Birge Hall, a large six-storey white building adjacent to LeConte Hall (the physics lab that produced six Nobel Prize winners). Was it the sun, the greenery or the students on their bicycles, that gave this campus a sort of holiday feel? "*I had a pretty relaxed lifestyle: at lunchtime I often picnicked by the Strawberry Canyon swimming pool and I would stop work early, around five in the afternoon*", he remembered. He would join Annie in their small apartment a stone's throw from the University, and they would go to the movies, a concert or a restaurant, or out with their new friends, like Phil Pincus, a student in the same laboratory as Pierre-Gilles — he would become one of his best friends — whose love of deliberately "*not always subtle*" puns hid a brilliant theoretical mind. "*Together, they were as thick as thieves*", jokes Annie de Gennes. At weekends, Pierre-Gilles and Annie would tour California in their convertible, treating the stay like a second honeymoon, even more enjoyable than the short week in Provence after their wedding five years before. *"I confess that I had a wonderful time in Berkeley. I missed the children a lot and wrote to them every day, but Pierre-Gilles and I were like young lovers. We were 25 years old and not a cloud in the sky! It was a time that really bound us together"*, explains Annie de Gennes.

"Although I felt that I wasn't working, the year turned out to be a fruitful one", recalled Pierre-Gilles de Gennes. The atmosphere in the lab was hard-working and productive, with Charles Kittel bossing everyone, to the point that students feared his explosions and nicknamed him "God". *"He could sometimes be harsh. He had asked one of his students to give a seminar. After twenty minutes, he commented stonily: 'You haven't grasped the subject. Do it again next week'"*, recalled Pierre-Gilles de Gennes. Merciless with his students, God treated the Frenchman with cordial benevolence, aware of his reputation as a promising young theorist.

Pierre-Gilles would prove himself worthy of the trust placed in him. In his very first seminar, he gained the respect of the physicists in the lab by introducing them to percolation. He published a first article in collaboration with Charles Kittel on "ferromagnetic resonance relaxation in yttrium iron garnets", recently discovered materials with interesting magnetic properties, then another, this time solo, in the prestigious *Physical Review Letters*, on the effects of the double-exchange mechanism in manganese oxides (A6-1). So the French physicist quickly found himself in the director's good books. *"When he had to be away for two weeks, Kittel chose Pierre-Gilles to run the classes in statistical physics in his place"*, reports Phil Pincus. Charles Kittel even went so far as to ask his advice. *"It might sound boastful, but I helped him with some question of paramagnetic resonance in a special case of rare earths. I expressed my view. Initially he was sceptical, but eventually he published an article based on my interpretation"*, reported Pierre-Gilles de Gennes.

The young Frenchman made a definitive mark with his American colleagues at a seminar given by Kei Yoshida, the lab's post-doc, on BCS theory. At the end of 1957, John Bardeen, Leon Cooper and John Robert Schrieffer published a revolutionary article explaining superconductivity, suggesting that the electrons were bound together in pairs, despite electrostatic repulsion (it is called the BCS theory from the initials of the authors). Wolfgang Pauli, the master of quantum mechanics, who had taught Pierre-Gilles at the École des Houches in 1953, was seated in the front row and suddenly asked a difficult question. The lecturer tried to answer, but was rebuffed by the master, and found himself lost for words. A heavy silence fell across the more than 200 people in the packed auditorium. Pierre-Gilles intervened: "*Perhaps I can clarify the situation. As I understand it,...*" Clearly satisfied with the explanation, Wolfgang Pauli signalled the speaker to resume. Pierre-Gilles had only wanted to help the young lecturer out of a tight spot, but to everyone in the room his intervention showed a profound — and unusual — grasp of the theory of superconductivity. "*I had understood neither the question nor the answer, like three quarters of the people there. It was only then that I realised just how good he was*", confesses Phil Pincus.

US Citizen?

His time as a post-doc was coming to an end. Bound by his military service obligations (he had been granted a deferment to the age of 27), Pierre-Gilles began to think about returning to France. "*Three months before I left, Mittel, who had become a friend, asked me to come and see him: "Stay here and become an American citizen. I'll offer you a position here and you won't have to do military service." I thought about it and discussed it with Annie. After a few days, I decided to say no. From a scientific point of view, it was tempting, but I couldn't get used to the idea of our children becoming American, which we would never be*", he confessed. Thinking that he was trying to up the ante, Charles Kittel offered him a much coveted tenured faculty position, which the Frenchman declined again, before regretfully leaving Berkeley (he would like to have extended his stay by a few months).

Charles Kittel's influence was visible not so much in the young man's research (de Gennes essentially developed his own ideas) as in his style. The American physicist was a great theorist of the practical. "*You have to do the calculations, then forget them and retain only the physics of the phenomenon"*, was the view he impressed on his students, demanding that they should keep equations out of their explanations. This approach suited Pierre-Gilles, who was already inclined to highlight mechanisms rather than formal description. After this experience, the young physicist would present his papers differently, adopting the more relaxed style of American researchers and sparing his audience the details of the calculations, instead focusing on the physical meaning of the results. This was in marked contrast with many French theorists, mired in formal approaches as if in an overtight suit, who would cover whole blackboards with equations, sending their audiences to sleep. Pierre-Gilles de Gennes would gradually become the champion of "hands-on physics", providing simple ("homely") explanations of the most complex mechanisms. For example, polymers, long molecules with complex physical properties, became spaghetti! It was a boon for the media, which became infatuated with him to the point that he became the TV face of physics after his Nobel prize in 1991.

Apart from Charles Kittel's impact on his style, these ten months on the American campus took Pierre-Gilles into the big league. He established a reputation with the Berkeley physicists that would spread across the whole country. He also built himself a strong network of American allies, headed by his close friend Phil Pincus. The same went for France, where he had just been awarded the Louis Ancel prize by the French Physics Society. It was his first prize... but it was his mother who was dispatched to collect it in his place.

At the end of September 1959, he took the long route back to Paris, giving several seminars on the East Coast, notably at the *General Electric* laboratories. There he met Charlie Bean, a physicist as enthusiastic as himself, who would become a friend. *"He was a very inspired physicist and a charismatic figure, even in the final years of his life. I remember a discussion with him about how ice skates work. We couldn't agree and he suddenly stood up and fetched some ice cubes from the fridge to show me that he was right!"* He also met a young Norwegian researcher, Ivar Giaever, who revealed an incredible result that he had just obtained: he had observed "walk-through-wall" electrons, capable of passing through an insulating layer by means of "quantum tunnelling" between a metal and a superconductor.[1] However, the result was supposed to be secret. Immediate alarm bells: the lab director, appalled at an unknown French visitor being in on the secret, summoned him to his office to swear him to silence. *"It was a major discovery (Ivar won the Nobel Prize for it in 1973), with potentially huge technological applications. So the director was very worried. The fact is that I was going back to Paris and would be catching the train for Brest the next day. I was joining the Navy, another world. So I didn't talk to anyone about it."*

Able Seaman de Gennes

Three days later, now an able seaman, he climbed aboard the *Richelieu training ship* in Brest harbour, a massive former World War II battleship. It was a big change. *"I found myself in the cold and wet, sanding down the deck of the Richelieu."* He was part of a group of young scientists recruited by the Navy as reserve officers, a privilege that kept them out of the Algerian War. Nonetheless, there was nothing comfortable about the conditions. *"We slept in hammocks slung one above the other in two rows, squashed together like sardines. When one person moved, the whole row swung. But I had got to know my neighbours, Jean Gavoret and Robert Goutte* (Author's note: they slept in alphabetical order)."[2] They were woken by a morning bugle call and took turns to perform the different chores. *"We frequently had to carry gigantic garbage containers off the boat"*, he recalled. *"But I don't have bad memories of those three months of training."*

The young man quickly got used to being woken in the small hours, folding and tidying his hammock, and putting on a tracksuit to run along the docks, before starting classes on nuclear propulsion,

---

[1] In classical mechanics, an insulating layer was considered an impermeable barrier, but not in quantum mechanics, as Louis de Broglie and Erwin Schrödinger predicted in the 1920s.

[2] Jean Gavoret would become director of the physics department at France's National Centre for Scientific Research, Robert Goutte director of the National Institute of Applied Sciences in Lyon.

submarine detection or... the different marine knots, etc. "*I was never bored.*" He also learned to navigate. "*I remember being ordered to manoeuvre a minesweeper in the middle of a flotilla of fishing boats. It wasn't easy, especially as minesweepers rock a lot and I was sick as a dog. I also had to steer a destroyer, off Brest: I was scared, because we had to sail between protruding rocks and the destroyer was travelling at 30 knots — a little over 30 mph. The slightest mistake would have spelt catastrophe! What's more, it was winter and I was freezing.*" Fortunately, everything went well, more than can be said of the parade ground drills. "*You damn well can't march in step!*", raged the drill sergeant. Instead of parading at the November 11 ceremony as planned, the young scientists were confined to the *Richelieu*. The drill sergeant had given up trying to teach them to march in time. They proved equally incapable of raising a mast or rowing a boat together. "*We were manoeuvring a sloop, and it was important to row together, otherwise we'd be in trouble. I'll always remember Pierre Cartier* (Author's note: a renowned mathematician), *who was often absent-minded, suddenly dropping his oars to note down an idea on a piece of paper, creating a huge mess!*", chuckled Pierre-Gilles de Gennes.

Cartier wasn't the only eccentric: Pierre-Gilles spent his little free time learning Russian, soon followed by the other trainees. He kept a language manual with him during long navy exercises. "*To practise, we used to speak to each other in Russian on the deck of the* Richelieu, *which caused a degree of panic, as our officers started to wonder if there weren't Russian spies on board*". Within two weeks, the young physicist had finished his manual and was now able to chat with some ease and produce a perfect rendering of the traditional Russian song, *The Volga Boatmen*. However, he had not learned the language to sing folk songs, but to read articles by Russian physicists.

Pancakes, Cider and Superconductivity

"*On furlough weekends, we were closely inspected before getting permission to leave. As we were a long way from Paris, we stayed in Brest in small groups*", explained Pierre-Gilles de Gennes. So he and two friends in the group, Jacques Joffrin and Julien Bok,[3] would rent rooms in a small hotel on Rue de Siam and he would spend his weekend... working. Having understood the BCS theory, he was now looking at the Russian school's approach, reading their articles in the original. Occasionally, he would leave his books for the evening to go and eat pancakes in the town, or to visit the surrounding villages. "*I can still remember us arriving in a cafe in some remote little hamlet, where the owner switched on the light while we were drinking and then switched it off again after we left. Some villages were still quite primitive*", he recalled.

On the Monday morning after furlough, the scientists would run along the docks to arrive in time for rollcall. The three months' training passed quickly and Pierre-Gilles, now a reserve officer, went to his posting, the Test Section, attached to the CEA's Military Applications Department in Paris.

Mushroom Cloud

Under its deceptive title, the Test Section was supervising preparations for the testing of France's first atomic bomb and all the experiments associated with it. Pierre-Gilles had not applied for this section, headed by Commander Kaufmant, a former naval officer.[4] His name had probably been put forward by Jacques Yvon or Yves Rocard, both involved in France's military nuclear programme, who would have vouched for the young man. Nonetheless, he was investigated to ensure that he had not, for example, signed the Stockholm Appeal against nuclear weapons.[5] This classified secret

[3] Jacques Joffrin would go on to become professor at Pierre et Marie Curie University. Julien Bok would become professor at the École normale supérieure and is currently emeritus professor at the ESPCI's solid-state physics laboratory.
[4] "*[The] Test Section was officially set up "from November 15, 1957", with the task of "preparing and executing all measures assigned to the CEA, on the occasion of nuclear explosions (...)". The memo setting it up, signed by the General Administrator and the High Commissioner, was, of course, "secret"*", "reports Jean-Pierre Ferrand, a naval officer seconded to the CEA, in a paper on the origins of the CEA's Military Applications Department.
[5] Research on the atomic bomb was conducted against the background of protest by supporters of Frédéric Joliot-Curie, instigator of the Stockholm Appeal.

defence section was not open to just anyone: you had to be squeaky clean.

January 1960, when Pierre-Gilles joined the Test Section, was a critical moment, since the first scheduled firing was just a month away. The atmosphere was feverish. *"And of course, I get myself sent to Reggane."* In February he was flown by military aircraft to the very place where the test was scheduled: the Saharan Military Experiment Centre near the oasis of Reggane, in the heart of the Algerian Sahara, an area a long way from the war but under very intense surveillance: almost 5000 troops were deployed there under the command of General Ailleret.

Despite the temperature of 130°C in the shade, the whole town had been built in the desert, with three-storey aluminium accommodation and laboratory buildings, and an airport to provide the crucial air bridge with Paris to bring in men and equipment. Pierre-Gilles suffered from the heat in the metal hut assigned to him, and envied the researchers in the two kilometres of cool laboratories dug into the cliff. It was here that the bomb and radioactive materials were stored. He would regularly leave his hut to prepare the measuring devices some thirty kilometres away at ground zero. He enjoyed driving around the desert, between the centre and the firing zone, despite the blazing sun. *"For the officers, there was a small swimming pool in the oasis: it felt wonderful to cool down in it from time to time"*, he recalled.

The test was imminent. The bomb was brought up to the surface, then moved with maximum precautions over the 60 kilometres between the base and ground zero. Once in place, it was hoisted on to a 300 foot high tower. Mannequins and surplus military equipment were placed all around, at varying distances, to test the destructive power of the bomb, with the measuring devices further away still.

On the big day, Saturday, February 13, 1960, Pierre-Gilles was in position, ready to start recording. *"My job was to do the measurements in the first thirty seconds after the explosion"*, he explained. Everything had been meticulously rehearsed. *"I had to switch on a certain number of recording devices, take photos and then take cover and wait for the light wave to go over and, before the sound wave arrived, switch off the devices so that they wouldn't get damaged.*" He was excited. A childhood memory came back to him. He was with his mother in Perros-Guirec, in August 1945, reading the newspaper on the hotel terrace. *"It says here that the Americans have exploded an anatomic bomb. What's that?",* he asked her. She corrected him: it was not an anatomic bomb, but an atomic bomb. Hiroshima had just been bombed. He didn't imagine that, 15 years later, he would be taking part in France's first nuclear test. As the time approached, there was mounting apprehension. *"Everything had been done in a big hurry and mistakes might have been made: for example, the bomb might have been five times more powerful than intended. That would've been a problem...",* he observed. Countdown. Time for him to do his job, but he had got the switches mixed up. The explosion was triggered. Transfixed, he couldn't take his gaze off the dazzling flash, then the great white ball that formed and gradually turned into a gigantic red, black and purple mushroom cloud. However many times he had seen pictures of such mushroom clouds, the spectacle unfolding before his eyes was so striking that he forgot to take cover... *"I did everything wrong, especially standing up and watching the explosion instead of lying flat... Then I felt the double shockwave, and boy did I feel it!"* Indeed, the shockwave was so powerful that his colleagues at the base, 60 kilometres from ground zero, felt their trousers flatten against their legs. "Gerboise bleue", France's first atomic bomb, had exploded. It would be the most powerful aerial nuclear explosion in the Sahara (its force was estimated to be three or four times that of the Hiroshima bomb).

"For the other explosions, I took cover", he stressed, aware of the dangers. As he was around 30 km from the explosion, he was probably not exposed to radiation. This was certainly not true of all the troops present, since in March 2009 the French government introduced a law to compensate the victims of nuclear tests conducted in Algeria between 1960 and 1966.[6]

---

[6] The documentary film "Gerboise bleue", screened in February 2009, tells the stories of the victims of France's nuclear tests in Algeria. Eleven Frenchmen involved in nuclear tests between 1960 and 1962 began court proceedings for involuntary homicide, supported by Aven (French nuclear test victims association).

The Generals' Putsch

It would not be the young physicist's only trip to Reggane. He was there, in particular, for the April 1961 test, which took place in unusual circumstances. On April 23, 1961, tension was high because of the putsch by the Generals (Maurice Challe, Edmond Jouhaud and André Zeller). *"I was very much against the generals, although I wasn't that bothered about the Algerian war, not enough — I was kind of mildly neutral"*, he confessed. There was a state of emergency on the base, and the troops were on the alert. A rumour was going the rounds that supporters of the Generals might fly down to Reggane and seize the bomb. *"The base was under military protection by more than 4000 men, commanded by General Ailleret. In principle it could hold out..."* However, might not some of the soldiers already in place be tempted to join the generals? There was a great deal of confusion. The fourth test was brought forward to prevent the bomb falling into their hands. But the temperature was too hot and the electronics failed. In addition, a thick sandstorm prevented optical measurements. This time, Pierre-Gilles was careful to protect himself, but the explosion released much less energy than expected: it was a fiasco. This would be the last surface explosion in the Sahara: subsequent tests from November 1961 onwards would take place underground, but Pierre-Gilles would not be there.

He was, however, at the Test Section's new centre in Bruyères-le-Châtel, forty kilometres or so south of Paris, investigating the impact of the bombs on the ground. For example, he was responsible for estimating how much force the ground could resist without deflagration taking place. The temperatures associated with underground explosions are so high that they vaporise all the surrounding rock. They emit shockwaves that spread at mind-blowing speeds of some 10,000 metres per second. *"I had read some fascinating literature on the subject, and on the bomb in general, in particular articles on shock theory written by Hans Bethe* (Author's note: who was responsible for the theoretical physics on the Manhattan Project and would be awarded the Nobel Prize in 1967)", reported Pierre-Gilles de Gennes. He talked to geologists and learned about soil mechanics.

Finally, he developed plasticity and elasticity models to assess the extent of the vaporised areas, depending on the power of the bomb, a model that would never be made public. He then compared his predictions with the computer simulations run by Jean Gavoret, his former hammock neighbour on the *Richelieu*, on the TERA10 super-computer, one of the world's most powerful computers, which belonged to the section. Initially impressed by the power of the instrument, he became disenchanted when he looked up previous simulations of spherical aboveground explosions. *"However refined they were, the simulations were not at all realistic"*, he complained. This was because they took no account of the ground, which is of primary importance: it acts like a mirror, in a way doubling the power of the source. Pierre-Gilles realised the limitations of computers and became somewhat distrustful of computer simulations. He would become a vocal critic of the uncritical use of these digital gas factories.

Land of the Rising Sun

Not all his time was spent at Bruyères-le-Châtel. Because there were links between the civil and military CEA, as a research engineer at the civil CEA, he had privileged access to the military side. This meant that although still on military service, he would go regularly to Saclay to pursue his research work. In April 1960, for example, he was given permission for a stay in the US, when he visited Duke University in North Carolina, meeting the superfluidity specialist Horst Meyer. He also went to the Oak Ridge National Laboratory where the first neutron reactor had been built for the Manhattan Project. He saw Charlie Bean at the *General Electric* lab, and also Harry Suhl — who had lectured in Grenoble — at *Bell Labs*, and then visited an IBM research facility. In all these places, he gave lectures on neutron scattering in magnetic media. Meeting more and more physicists, he gradually extended his network of contacts. His American trip lasted two weeks.

"In September 1961, I was also given special leave to attend the International Congress of Theoretical Physics in Japan. I arrived in Kyoto, a

city that immediately captivated me and that I still love, where my host was the Yukawa Institute. What was strange was that the Institute was named after Hideki Yukawa, 1949 Nobel Prize for physics, who was still alive and welcomed me to the Institute in person. We wouldn't do that in France", he remarked. Pierre-Gilles took a real shine to Japan. *"I fell in love with the country and I feel affinities with their culture. I became accustomed to their way of not arguing, of laughing when embarrassed... I spent hours sketching in the temples."* Although communication was sometimes difficult, he formed contacts with a number of Japanese physicists, in particular Ryogo Kubo, a specialist in phase transitions, whom he had already met at the Les Houches School. Impressed by the young man's intelligence, Kubo would invite him to give regular lectures in Japan, becoming not only his principal host, but also a primary contact, especially when the French physicist began his work on polymers.

During his military service, he was so free of restrictions that he became research supervisor for his first doctoral student at Saclay, Françoise Hartmann-Boutron. He had a crew cut and was somewhat patronising: *"I'm going to give you a calculation to do. I expect it will take you at least three months"*, he commented. Irritated, she set to work and won his respect by returning it eight days later. *"She is one of those people who follows things through to the end and doesn't give up when the going gets tough"*, he approved. He saw her every week, giving her a few instructions or calculations to do. He had her working on yttrium-iron garnets (YIG). *"These days, they are standard materials in microwave electronics, but at the time they had just been discovered and they raised questions about spin dynamics. When I was at Berkeley in 1959, Charles Kittel and I tried to interpret so-called ferromagnetic resonance, but our article was a complete turkey, because it ignored certain important interactions"*, he acknowledged. So he formulated a new description that included Suhl interactions, and demonstrated the existence of collective modes (spin waves), which had previously been ruled out (A6-2).

Here, once again, he missed out on a discovery, for the second time in his short career, after the case of spiral magnetism. *"We were looking at how spin waves pass through a Bloch wall* (Author's note: the boundary between Weiss domains, regions with the same magnetic orientation) *and how the wall itself oscillates, which was quite bizarre. And we discovered strange properties in the scattering matrix* (Author's note: which describes the transformation of the incident wave into a scattered wave). *If we had been a bit cleverer, we could have come up with soliton theory"*, he acknowledged. The theory of solitons in magnetic materials would be developed four years later, in 1965:[7] solitons are solitary waves that cause the spins of a magnetic material to switch as they pass, propagating without changing shape. They resemble tidal bores, the single waves that travel up river against the current (first studied by the Scotsman John Scott Russell on a canal in 1834). Was it lack of time or lack of self-confidence? In any case he missed out.

Pierre-Gilles did not waste his time during military service. At the end of the required 27 months, apart from the secret research he conducted on the effects of underground explosions, he had published several new articles. That was not all. *"I was lucky enough to be able to go home every evening and to have enough time to study superconductivity"*, he explained. In fact, since 1956, he had spent every free moment delving into the subject in his little study on Rue du Mail, assimilating the BCS theory and everything that the Russian school had come up with on superconductivity. He now had an equal grasp of both approaches. He had some ideas of his own, but would he be able to develop them at the CEA? The question would not arise, because Pierre Aigrain, at a chance meeting outside the Sorbonne, told him the good news: *"That's it! You've been awarded a lectureship at Orsay* (Author's note: the equivalent of a professorship today)! He was not yet 30, and the gates of academia were opening before him.

---

[7] It was discovered by Norman Zabusky of *Bell Labs* and Martin Kruskal of Princeton University. The idea had been introduced in 1940 by Rudolf Peierls (to describe the change in the spins).

Chapter 7

Team Leader

One morning in September 1961, Pierre-Gilles de Gennes drove carefully around the Orsay campus construction site. Buildings were springing out of the ground like mushrooms, a sign of the rebirth of physics research in France. He parked in front of the recently completed Solid-State Physics Laboratory (LPS), where he has been assigned. *"The lab was created in 1959, thanks to Yves Rocard, who had set aside some space for us in the premises he was having built around the École Normale's linear accelerator at Orsay"*, recalls its director Jacques Friedel.[1] Aware that Friedel wanted a research lab, the Director of the École Normale Supérieure had looked at the plans of the future accelerator building, and sketched an extra rectangle with the words: *"Here's a building for solid-state physics!"* The LPS — which would enable Jacques Friedel to claw back France's twenty-year

---

[1] Jacques Friedel had brought in two experienced physicists, André Guinier and Raimond Castaing. At 50, André Guinier was the oldest of the three men (he was behind the design of a very powerful x-ray diffraction chamber). Raimond Castaing, his student, invented the electron microprobe which could show matter at the micrometric scale with the x-rays emitted when a sample is bombarded with an electron beam.

arrears on the US and the UK in solid-state physics — thus came into being at the stroke of a pen.[2]

The young lecturer strode energetically towards the building. Unbeknownst to him, his appointment had caused something of a furore. Philippe Nozières, his classmate and the top student in their year at Rue d'Ulm, had been a candidate for the same position. After completing his thesis in the US,[3] Nozières had accepted a heavy teaching load at Orsay, hoping that his efforts would be rewarded with a lectureship, only to see it go to his friend! What made it all the harder was that it was rumoured that Pierre-Gilles had threatened to settle in the US if he did not get the post. The rumour was unfounded, but it soured their friendship for a while and gave an extra edge to their previously good-natured rivalry.

The appointment also raised hackles in Grenoble, where the physicists took a dim view of the young theorist's encroaching on their turf. *"Superconductivity, along with low temperatures, was a speciality of Grenoble.* (Author's Note: Superconductivity appears at very low temperatures, below −250°C, a process that required expertise which was rare at the time). *So they were not that pleased to see me turn up... As they had a lot of influence in the committees, we didn't get any CNRS money"*, recalled Pierre-Gilles de Gennes. Fortunately, he had US Air Force funding. *"It wasn't a huge amount, but it gave me independence from the committees. That's why the contract with the U.S. Air Force was important."*

The young lecturer pushed open the glass door of the building and took the staircase to his small first-floor office four steps at a time. He laid his briefcase on the table and rubbed his hands. Things were looking good. He had only one teaching commitment, a postgraduate course on solid-state physics, which meant no lecture halls crammed with first-year students — no bad thing. On the

---

[2] At the time, there was no solid-state physics in France other than Louis Néel's work on magnetism in Grenoble (based on a classical approach) and the semiconductor research headed by Pierre Aigrain at the École Normale Supérieure.

[3] He had developed a complete theory of the low-temperature properties of free electron gases to describe metals. His supervisor, David Pines, said of it that the student had surpassed his master.

research side, he at last had free rein. His goal, to create a team to study... superconductivity, of course! All those years of work would not go to waste. So what if the competition was intense. Like David facing Goliath, he was no more afraid of starting a lab from scratch than of rubbing shoulders with the world's top American and Russian solid-state physicists.

Mystery of Superconductivity

The fact was that Pierre-Gilles was tackling the hottest topic in the field, as the mystery that had always surrounded superconductivity was beginning to dissipate following the emergence of the BCS theory in 1957. *"The phenomenon was discovered in 1911, but for a long time no one understood it."* The new laws of quantum mechanics gave a marvellous description of everything occurring in metals, but seemed unable to explain why electrons circulate without resistance in those same materials at very low temperature. *"Back then, once started, a current would flow through the metal for a year. These days, it could circulate for a million years."* Even the great Albert Einstein considered the problem, but came up against a brick wall, like every other physicist. Superconductivity became even more mysterious in 1933, when a German physicist called Walther Meissner showed that there is no magnetic field inside a superconductor. Double whammy: superconductivity was no longer just characterised by resistance-free electrical conduction, but also by the disappearance of the magnetic field. This raised a new question: which of the two effects caused the other? *"The guy who then understood superconductivity best, but was largely ignored at the time, was Fritz London"*, opined Pierre-Gilles de Gennes. The German physicist had shown that superfluidity in helium 4 (a phenomenon where the helium, cooled to below −270°C, flows without friction) was caused by a Bose-Einstein condensate, a very low-temperature phase in which all the particles are in the same state, with no dissipation of energy (which explains why superfluid helium flows without friction). This gave him the idea that electrons in a superconductor might be precipitated into a similar state and flow without resistance. The one snag: the only particles capable of condensing in this way are bosons (such as helium atoms).[4] However, electrons are not bosons and hence in principle cannot condense. Attractive as it was, the scenario did not stand up (A7-1).

In the 1930s, a brilliant Russian theorist, Lev Landau, took a totally different approach to superconductivity. Lev Landau was the author of a general theory of phase transitions, where an "order parameter" — which in a sense represents the "degree of order" in a system — is cancelled in the transition to disordered (higher temperature) phases: for example, the order parameter of iron is cancelled in magnetisation, the transition from the ferromagnetic to the paramagnetic phase (A7-2). The Russian physicist applied his theory to superconducting transition, using as his order parameter the intensity of the superconducting currents (which is effectively cancelled in the metallic phase). His first attempt failed, but then, after the war and following a year in one of Stalin's prison camps, he and Vitaly Ginzburg developed a theory which for the first time correctly described the superconducting transition. However, it said nothing at all about the mechanisms involved! So what objects are characterised by the order parameter? Certainly not electrons, which are not bosons. The question was left hanging...

Although incomplete, the Ginzburg-Landau theory of 1950 marked a major advance, of which US physicists would only learn years later: at the height of the Cold War, the New York dockers, in a fit of McCarthyist fervour, threw shipments from the USSR into the harbour, including... Soviet physics journals. It would be a long time before Lev Landau and Vitaly Ginzburg's ideas would penetrate the Iron Curtain separating the Eastern and Western blocs.

For their part, Russian physicists — in particular one Alexei Abrikosov — continued to follow the trail blazed by their two colleagues. To understand his work, one needs to know that all superconductors return to normal when subjected to a strong external magnetic field. Using the Ginzburg-Landau equations, the Russian

---

[4] All particles are classified as bosons or fermions, depending on their spin. Particles with zero or whole spin are bosons, e.g. photons, helium atoms, etc. The others are fermions (electrons, neutrons, etc.).

physicist showed that so-called type II superconductors, when subjected to a strengthening magnetic field, gradually lose their superconductivity as the field value rises from a value of HC1 to HC2 (losing it completely at field values in excess of HC2) (A7-3).

During this time in the US, physicists showed that superconductivity not only involves electrons, but also interactions between electrons and phonons (collective vibrations of the crystal lattice). But however they manipulated the equations, they got nowhere. In 1956, at the international theoretical physics conference in Seattle, Richard Feynman, the future Nobel Prize winner so admired by Pierre-Gilles, would acknowledge the general failure: *"We haven't got enough imagination."* Yet there was light at the end of the tunnel. An Australian physicist, Max Robert Schafroth, put forward a hypothesis that would solve the problem: while it is true that individual electrons are not bosons, two paired electrons are. Could the condensate be made up of pairs of electrons? *"In 1957, a brilliant young student called Leon Cooper had an amazing idea: he looked at two electrons, in the presence of a sea of electrons, which form a linked state that is quite fragile and strange. The student teamed up with John Bardeen, an old-timer who knew his quantum physics, and John Schrieffer, who had strong calculation skills, and between the three of them, in six months, they developed the complete BCS theory"*, explained Pierre-Gilles de Gennes. They discovered the mechanism responsible for superconductivity (which would win them the Nobel prize in 1972): despite Coulomb repulsion, the electrons in a superconductor join together in pairs, but they are very "elastic" pairs, since the electrons are only linked through minute deformations of the ion lattice (A7-4). The Russian physicists then understood that the mysterious objects described by the order parameter in the Ginzburg-Landau equations were not electrons, but pairs of electrons! Lev Petrovich Gor'kov, Landau's student, then reformulated the equations using the microscopic description of BCS theory.

Entering the Golden Age

So Pierre-Gilles was joining the race at this critical juncture, just after the discovery of the microscopic mechanism responsible for superconductivity. He was well armed and full of ideas, having studied the subject in his moments of leisure, as others might collect stamps. *"I can still see myself in my study in Rue du Mail, before the BCS theory was published, imagining that electrons left a wake because they excited the phonons in the metal. In fact, I had considered that kind of stuff, but it hadn't led anywhere. I was also trying to understand why there was a gap, a forbidden zone, in the excitations... one of the big mysteries of superconductivity that I'd always wondered about. I followed everything, I read everything in detail."* From this point, he would find his way through by using both the BCS theory, which describes the phenomena at microscopic level, and the Ginzburg-Landau equations, which take account of the geometry of the samples — we know that, unlike most western physicists, he was familiar with the Russians' work. In practical terms, he began by simplifying the Ginzburg-Landau equations — *"I tried to convert what the Russians had done into simpler language, because their mathematical language is often rather clumsy"* — and applying them in different situations. And secondly, he generalised the BCS theory using the more digestible version established by Nikolaï Bogoliubov in 1959 — *"the initial, let us say "unexpurgated" formalism of BCS was pretty obscure"*, he explained — to non-homogeneous systems. This led him to equations that would be called the Bogoliubov-de Gennes equations — *"They are mainly his, all I did was write them down"*, he claimed.

A Different Kind of Lab

To test his ideas, Pierre-Gilles wanted to set up a proper experimental laboratory, not a school of theorists. The first experimentalist he recruited was Étienne Guyon, a student of André Guinier. He invited him to his office and immediately adopted the familiar French *tu* form of address, as he did with all his young colleagues.

— *"Are you familiar with superconductivity?"*, he asked.
— *"I attended Bardeen's courses, at Illinois University"*, the student replied.
— *"So you must know a fair bit"*.
— *"It might be good to go over it again...."*

Pierre-Gilles smiled, knowing that John Bardeen, though a brilliant researcher, was a lousy teacher, with a habit of mumbling and wiping his equations off the board the moment he had written them down. He immediately delivered a crash course in superconductivity, frequently jumping up to draw a diagram or scribble an equation on the board, which by the end of the afternoon was covered in symbols and drawings. *"Nothing in science has ever had as much impact on me as what I heard that afternoon"*, asserts Étienne Guyon. The student had found his master (to whom he would remain forever devoted). Pierre-Gilles then led him up to another floor and threw open the door. *"Here is the superconductor physics lab"*, he announced, pointing to the large... and empty... room.

The second experimentalist he enrolled was Alexis Martinet, a student who had left Marseille University on learning that a superconductor research group was being set up at Orsay. *"This is a lad who knows what he wants"*, the theorist thought. *"From the moment you met Pierre-Gilles de Gennes, you knew that this wasn't just anyone, although he was not intimidating... He went straight to the point and explained that he wanted to bring quantum tunnelling to France as a method of superconductor research — a very powerful experimental technique for studying superconductivity. He immediately realised that he had to have this instrument in his lab. So Étienne and I collected the equipment to mount the apparatus"*, recalls Alexis Martinet.[5] Pierre-Gilles had given his two apprentice researchers a tough task, but he made sure that they were under the protection of Pierre Perio. *"He was a former paratrooper, who had switched to crystallography at CEA Saclay. He took a shine to me, though he could be tough with researchers who were too arrogant or sure of themselves. He became a sort of uncle figure for me"*, admitted Pierre-Gilles de Gennes.

The team then expanded to include Jean-Paul Burger, holder of a PhD from Strasbourg University, who had been assigned to the LPS (solid-state physics lab) for his military service, and Guy Deutscher, a

---

[5] This technique was based on the discovery by Ivar Giaever, who disclosed it to Pierre-Gilles de Gennes during the latter's visit to *General Electric* after his post-doc year in 1959.

doctoral student seconded by *Bull* to the LPS. Pierre-Gilles nick-named these four young scientists his "musketeers": Étienne Guyon, permanently wired, boiling with ideas; Guy Deutscher, the planner, who thought everything out before starting an experiment; Alexis Martinet, a peerless experimentalist; and Jean-Paul Burger, the moderator for the group. In addition, there were two other theorists in the Orsay superconductor group: Christiane Caroli, a former student of Jacques Friedel (she would become one of France's most brilliant female physicists), and Jean Matricon, another graduate of the Ecole Normale and Pierre-Gilles' "twin" (they were born on the same day).

Small Lab, Big Future

After installing the equipment, the musketeers made their first samples — "tunnel junctions", very thin sandwiches of metal, oxide, insulator and superconductor. But after a year of trying, the sought-after tunnel effect had not appeared. Whispers began to go around: *"It was madness to go into very low temperatures with such a small team."* Temperatures of less than −250°C were required, no picnic! The musketeers persevered, changed the experimental settings... *"We knew it was possible, but we couldn't do it. So we played around with the vacuum conditions, the preparation... We tried everything"*, recalls Alexis Martinet. Nothing worked. They were starting to get discouraged: apparently there were tiny holes in the samples... Pierre-Gilles became worried. Was he taking his students down a blind alley? The situation would soon resolve itself, following a serendipitous series of events.

It began with an article that Étienne Guyon gave Pierre-Gilles to read.[6] *"The article described a surprising experiment. It involved superimposing two thin layers of metal, one lead (the superconductor) and the other copper (normal metal). The article said that as more copper was "added", the temperature at which the lead became superconducting — which is 7K in a normal sample of lead — fell"*, explains Étienne Guyon. A few days

---

[6] Article by the German physicist P. Hilsch (published in *Z Physik* in 1962).

later, Pierre-Gilles had already come up with an interpretation of this "proximity effect", demonstrating that it was not caused by the appearance of a parasitical intermediate compound, as some thought, but by the diffusion of Cooper pairs from the superconductor to the metal. He showed Étienne Guyon his article, under both their names.

— *"But I didn't do anything"*, protested the student, glancing at the abstruse lines of calculations.
— *"Yes you did. You brought the experiment to my attention"*.

This article, the basis of the "de Gennes-Guyon-Werthamer" theory which predicts the temperature of superconducting transition from types of metal in a superconductor/metal sandwich), became a key source. Étienne Guyon was invited to present it at a major superconductivity lab in Cambridge.[7] *"Pierre-Gilles de Gennes was always very generous with young scientists. He gave them his trust and sent them to give papers at conferences or at other labs. Over and over again, young researchers would find themselves associated with a piece of research, or cited, without necessarily deserving it... He was always like that"*, commented Étienne Guyon. During his visit to Cambridge, the student told his hosts about their problems with the experiment. Their sympathy aroused, they passed on a tip: when the metal oxidises, you have to set off a discharge in the vacuum chamber to produce an insulating layer of uniform thickness. It was an idea that the musketeers would never have thought of on their own! Étienne Guyon was in a hurry to get back to test this new *modus operandi.*

There were still technical problems to be solved, but the group was filled with renewed hope, and on one particular afternoon Alexis Martinet observed the experiment with a different eye. He

---

[7] This lab was the Royal Society Mond Laboratory, where Brian Josephson was doing his PhD under the supervision of Brian Pippard and John Adkins (Josephson was a physicist who showed Cooper tunnelling between two superconductors separated by a "bridge" (a layer of insulation or metal), which won him the Nobel Prize in 1973).

fixed his gaze on the tip of the needle that recorded the current's response. So far, the lines had remained obstinately straight, indicating a standard metal. This time too, there was no movement of the tip as the paper rotated, a sign that the current was not penetrating the sandwich. *"Then, suddenly, I saw the tip jump! The line became a curve, the current was getting through!"*, remembers Alexis Martinet. The young man could not believe his eyes. He picked up the phone to call Pierre-Gilles, who immediately joined him. They ran the experiment again, and there was no doubt, what they were seeing was a tunnel current. *"He was so pleased and relieved! At last, the solution was within reach. We spent quite a while just looking at the result"*, recalls Alexis Martinet. After a year of failed experiments, work could at last begin.

During this time, Guy Deutscher and Jean-Paul Burger had set up another experiment to measure magnetic susceptibility, i.e. a material's response to a magnetic field. With their setup, they could track the penetration of a magnetic field into a sample to the nearest nanometre. When they conducted the experiment on a bimetallic superconductor, they discovered that the metal layer of the bimetal seems to screen the applied magnetic field, as if it were subject to the Meissner effect, which would mean that the metal had become a superconductor by contact with the superconductor. Though its superconductivity disappeared when the applied magnetic field increased this only happened at a level called the "breakdown field", which was lower than the usual critical field. Pierre-Gilles was very surprised. He had not at all predicted this effect, even describing it as *"improbable"*. True, he had established that Cooper pairs could diffuse from the superconductor to the metal, but between that and the metal itself becoming a superconductor, there was a huge step! *"Here he had come up against a conceptual difficulty on the very cause of the Meissner effect"*, opined Guy Deutscher. And he would not allow the result to be published until he had understood it. The weeks passed...

It was only when Étienne Guyon and Alexis Martinet demonstrated the existence of the "breakdown field" caused by quantum tunnelling, that he would realise that the metal in the sandwich genuinely becomes superconducting through the proximity effect. A few days later, he turned up with a big smile and... an explanation, which would constitute one of his major contributions to superconductor physics. The musketeers confirmed his interpretation and results came thick and fast between 1962 and 1967.

Cowboys in Utah

One day, when Daniel Cribier, his Saclay classmate, was dropping him off at Orly Airport, Pierre-Gilles de Gennes handed him a sheet of paper: *"This is what I'm planning to talk about tomorrow in the US. The neutron physicists will fall over themselves to do the experiment."* A few lines on the sheet revealed that the "vortices" (A7-3) present in type II superconductors could be revealed through neutron scattering, but would be very hard to see, because he and Jean Matricon had estimated that the level of scattering would be 1000 times weaker than usual. *"Hard, but not impossible"*, thought Daniel Cribier, who quickly began the experiments with Bernard Jacrot. Together, the two worked for more than six months, becoming the first people in the world to use neutron scattering to observe a vortex lattice, trouncing four rival American teams in the process. They phoned Pierre-Gilles at his faculty office to announce the good news. *"This I have to see"*, he responded, and dropped what he was doing to go and admire the result.

With this success in the bag, Pierre-Gilles, Daniel Cribier and Jean Matricon received a royal welcome in the United States, first in Brookhaven, site of the world's most powerful neutron reactors, then at the Bell labs in New Jersey. *"We're impressed at the results you've managed to achieve in France"*, they were told. After a triumphal tour, they allow themselves a week's holiday in Utah and fulfilled a childhood dream, temporarily dropping their equations to... play cowboys. They bought jeans and cowboy hats at a drugstore and set off... across the desert on horseback. Then, they took in the Grand Canyon via the Nevada desert in a rental car. During the drive, they were chased and stopped by the police because Daniel Cribier was speeding. The cops came running over, guns in hands, shouting: *"One man only."* Crestfallen, Daniel Cribier got out of the car, while his accomplices stayed in the car, killing themselves with laughter!

Back in the lab, Pierre-Gilles continued his exploration of super-conductivity, this time working with films, to see what happens when a strong magnetic field is applied to a type 2 superconducting film. In the bulk state, the superconductor gradually loses its superconductivity (between HC1 and HC2), but what happens with a film? The theoretician called on Daniel Saint-James, a physicist at Saclay. *"One day, he said to me: 'Why don't you see what happens when we solve the Ginzburg-Landau equations?' I spent a fortnight writing the program and doing the calculations with perforated cards on a computer that was very powerful for its time, which filled a whole room, and I found something odd…"*, reported Daniel Saint-James.

He called Pierre-Gilles de Gennes to his office to see the results. As expected, superconductivity did disappear in the sample as the magnetic field was increased, but there was something that bothered the Saclay physicist.

— *"Look, it's strange. It is as if a sheet of superconducting material remains at the surface, even when a very strong magnetic field is applied"*, he pointed out.

Pierre-Gilles considered the anomaly. Normally, with a strong magnetic field, the superconductivity disappears, whereas in this case it was still present at the surface. He rubbed his chin and thought, pacing around the room.

— *"It must be linked with the boundary conditions, he mused. The wave function of the electron pairs can be reflected at the interface with the air, unlike at the interface with a metal"*.

Since this hypothesis opened up new possibilities, they then tried to solve the Landau-Ginzburg equations directly rather than by computer, and discovered that there exists a third critical field, which they named HC3, up to which superconductivity persists (its value is 1.69 times the HC2 field) (A7-7). *"No sooner had the result been published than I got a call from Phil Anderson, grand master of theory at Bell Labs: our discovery had not gone unnoticed! I knew that we had pulled off a coup"*, recalled Pierre-Gilles de Gennes with a smile. *"So I then turned to my young experimentalists and said: "There must be a special phenomenon, a sort of superconducting sheath that occurs in certain conditions: how can we get to see it?" Clever as they were, they quickly devised an experiment that revealed the HC3 field and measured that it had the expected value."* There was general rejoicing. With this experiment, the Orsay superconductor group achieved an international reputation. The group expanded, and from 1965 included a dozen doctoral researchers.[8]

Experimentalists and Theorists

The credit for this success lay with Pierre-Gilles, who empowered his team to an unusual degree. For example, he treated his students as colleagues and allowed them to organise their work as they wanted rather than assigning them specific tasks, as did most laboratory chiefs. In fact, tasks were shared: one day, one of the musketeers would concentrate on calculation, whilst another prepared samples, another did measurements and the fourth explored the literature. The next day, the first would go through the results while the second emptied the oil pumps, etc. Of course, there were sometimes arguments — for example, accusations of skipped chores, etc. — but the conflicts never went far and, in any case, never reached the boss's ears.

Above all, Pierre-Gilles managed to get theorists and experimentalists to work together. *"Usually in labs, either the theorists or the experimentalists move faster, and they get out of step. It's rare to advance at the same rate. A theoretical team can have three ideas a day, each of which requires a year of work. Conversely, if the experimentalists hit a problem, by the time the theorists solve it, the others have moved on to something else. As a result, theorists and experimentalists often work separately"*, explains Alexis Martinet. This is what Pierre-Gilles wanted to avoid, by

---

[8] Amongst them Michel Cyrot, Jean-Pierre Hurault, Santi Mauro, etc.

encouraging his theory students to get involved, as he himself did, in setting up experiments — he would often surprise people by his remarkable grasp of detail, such as the type of weld likely to distort results — and by trying to entice them out of their formalist prison. *"I had been through École Normale Supérieure and I knew how to crunch integrals, but in fact I knew nothing about physics. I remember once telling Pierre-Gilles de Gennes that some problem was insoluble, because the mathematical function was impossible to integrate. He replied with one word: "Out!" He was right: that's not how a physicist should think. I'm infinitely grateful for the training he gave me. He worked without preconceived ideas — for him, techniques were no more than tools"*, recalls Christiane Caroli. On the other side, he encouraged the experimentalists to look at the equations. *"He would often turn up in our office, take a sheet out of his old briefcase and say: "I've done a little calculation. See what you can do with this." He expected us to understand his calculations and devise an experiment to verify them. He would begin by making us look at the calculations, and only discuss them after we had done so"*, relates Guy Deutscher.

"Pierre-Gilles de Gennes did a lot to get theorists and experimentalists to talk to each other. He set an example and showed that it is worth doing. There are not many places in the world where experimentalists and theorists were as close", assesses Jacques Friedel. The team spirit was such that they wrote joint articles and signed them "Orsay superconductor group" — unheard-of! To begin with, nobody objected, then some began to complain. *"It was a kind of idealism on his part. It was nice, but in fact everyone knew that Pierre-Gilles de Gennes was behind the group, which prevented the others building a reputation. He later realised this and abandoned the idea"*, recalls Jacques Friedel.

Although he treated his students as equals and gave them his complete confidence, Pierre-Gilles was still the boss, and not always the mildest of bosses. *"In the early days, he could be very sharp. Later on, he mellowed a lot..."*, recalls one of his collaborators. His students would think twice before asking a question, for fear of being mercilessly put down if it was not relevant. Some students were so in awe of him that Phil Pincus, his Berkeley friend, nicknamed him "the ogre". Despite this, aware that they were working alongside a physicist of a remarkable kind, they were eager to give of their best.

"Backscratchers" Club

From the 1960s, Pierre-Gilles began to build a reputation on the international stage. At the Ampère conference in Eindhoven in the Netherlands in 1962, for example, he gave a lecture that attracted the attention of Hendrik Casimir, Scientific Director of Philips — one of the big guns in physics — who would help to establish his position in Europe. In the USA, he already had his own network, headed by Phil Pincus, who had taken on the task of "promoting" him. *"But it's fair to say that it was with his results on superconductivity that he really hit the big time"*, reckons Jacques Friedel. *"Before his research on superconductivity, he was the student everyone was talking about, but then he became a young grand master. In the 1960s, he was already a big name in physics"*, agrees a female physicist. He was invited to give lectures everywhere in the world,[9] travelling to the US several times a year. He spent time in Japan and Cambodia in 1965, then in Venezuela in 1966 to give classes in a summer school on superconductivity. *"I made friends with an American physicist called Laurence Mittag — he reminded me of Ernest Hemingway with his white beard. The two of us used to explore the streets and bars of Caracas at night"*, he reminisced. From every corner of the planet, people sought his opinion or an invitation to Orsay — he was careful to reply to them all and keen to welcome foreign visitors, encouraging his students to associate with them. *"Pierre-Gilles became very popular, which attracted a host of visitors to the LPS, including some of the best-known physicists in our field, figures like Michael Tinkham, Kazumi Maki, Charles Franck or Brian Pippard. Orsay became a favourite destination for the Americans"*, recalls Jacques Friedel.

Pierre-Gilles had made a name for himself, but also wanted to give credit to his collaborators, whom he was always careful to cite in lectures, to the point that, for a joke, Daniel Saint-James and his colleague Gobalakichena Sarma created a "backscratchers club" named the 'Société de la brosse à reluire mutuelle' (SABRM) — mutual

---

[9] Or almost: he would refuse to set foot in the USSR for ideological reasons and would send colleagues in his place. Similarly, he would never go to Spain under the Franco regime.

backscratching society — appointing Pierre-Gilles as its honorary chairman. The members of this club, which included the other musketeers, numerous physicists from Saclay, etc., were obliged to mention the other members in all their lectures, even if they hadn't contributed to the results!

Demanding Teacher

Apart from research and conferences abroad, Pierre-Gilles also devoted a significant proportion of his time to teaching. "*I prepared my classes carefully. I hardly ever used textbooks — in fact there were very few of them — but worked directly from scientific articles*", he explained. His postgraduate classes at Orsay began early in the morning.[10] He would stride into the room, throw his old leather briefcase on the small desk on the dais, and go straight to the blackboard, without a word. Then he would introduce the subject of the class in a quiet voice, chewing on a pipe, writing the title and lesson plan on the board. The hubbub in the room would immediately die down. The students knew that the essence of the class would be conveyed in these first five minutes. Too bad for latecomers! Then, he would put down his pipe, a sign that the real business was about to begin. He would go straight into the subject and continue non-stop, striding from one end of the podium to the other. He would draw a graph on the board, describe the classical perspective on the situation and then, with a slight smile, announce: "*In the quantum world, it's the opposite.*" He would pause for effect, maintain the suspense for a moment, then continue: "*Indeed, if we just think about it...*" The students would hang on his every word, every sentence had its significance. Indeed, Pierre-Gilles would describe the hypotheses and comment on the results without going through the intermediate calculations. As soon as he finished his class, often ahead of time, he would pick up his briefcase and leave. When the

---

[10] He also gave classes for the young researchers at Orsay, in the postgraduate course on atomic physics headed by Jean Brossel at the ENS and at ESPCI (Paris Higher School of Industrial Physics and Chemistry).

students left the room in their turn, they would feel they had understood everything, but a few days later, they would often find that their notes no longer made sense. *"But then we would get together and reconstruct the lecture"*, reports Liliane Léger, one of his pupils, who would go on to do research with him. *"We were motivated, because what he taught was new compared with what we had previously learned: they were results that might have been obtained just 10 years earlier. We felt we were part of an adventure, involved in the progress of science. For us, Pierre-Gilles de Gennes was something of an idol"*, she recalls. Particularly testing were the end of year oral exams, because he did not confine his questions to the content of the lectures, but would ask new ones, for example: *"A vacuum is gradually created in a sealed room containing a microphone. At what stage will the sound cease to be audible?"* Even the best students were intimidated, in particular those from France's elite Ecole Polytechnique, on whom he was particularly tough. *"They tended to be know-it-alls, who thought they were the centre of the universe"*, he declared candidly. Many contemporary solid-state physicists went through his hands. *"I kept a notebook with the names of every student and a comment on their answers. There were "pluses", "minuses" and exclamation marks when it was really garbage, and then the final grade. It's funny, because a lot of the students became university teachers"*, he grinned.

Head of the Family

Despite his research, his teaching and his travels, the scientist maintained a family life. When he was awarded his university appointment, he bought a large, handsome townhouse, a quarter of an hour's walk from his lab. In both Paris and Orsay, Pierre-Gilles always left all the day-to-day stuff to his wife. Nothing was allowed to distract him from research. He would get home quite late, but always in time to eat with the children, now aged four, six and eight. *"In the evening, at mealtimes, he got into the habit of telling us about the films he had seen with our mother — they used to go about once a week. Sometimes, he would also talk about new experiments at the lab"*, recalls his youngest daughter Marie-Christine. *"Our home culture was mainly oral. We had no TV and comics were forbidden"*, reports his elder daughter Dominique. After the meal, he would read and listen to music or the radio. *"For example, they used to follow* Les maîtres du mystère [Masters of mystery] *on an old wireless set. We would hide in the corridor, flat on our stomach, to listen as well"*, she smiles. *"Around 10 p.m., I would make him a small round of sandwiches. He always ate like a horse"*, notes Annie de Gennes.

Though he never worked in the evening, at weekends he would shut himself in the study that Annie had set up for him — as she had in the Rue du Mail apartment — even installing insulated panels to minimise noise. The children learned not to disturb him. *"We were very free in the house: we could shout, sing and do what we wanted, provided it wasn't near daddy's office. There was always absolute calm near the place where he worked"*, recalls Dominique. The children would sometimes open the study door to show their friends *"a daddy working"*.

When their father emerged from his den, he did not need to raise his voice to be obeyed, such was his natural authority. Even when he gave them chores that they hated, such as wrapping the pears on the tree to protect them from insects the children followed his orders without demur. He rarely smacked them, except in the case of very serious misbehaviour. And, however often his son Christian, who was a bit of a daredevil, tried to jump off the terrace or raced down the slopes in his toy car, his father never said a word, seeing these games as *"somewhat extreme experiments"*.

He showed little interest in the children's grades or homework assignments. Just once, he tried to help Christian with his maths exercises: he sat in a corner with his guitar waiting for the child to solve the problem, but the poor boy was completely stuck. So Pierre-Gilles decided to call it a day. *"He would sometimes start explaining things in maths and physics, but if he saw that we couldn't understand, he would sigh that we were dunces and that was that"*, recalls Dominique. He preferred to spend the little time he had with them reading aloud or playing. For example, he would join them in a field next to the house to launch water rockets. And sometimes, he would play table football or go pigeon shooting with them. When he came back from conferences abroad, he would often bring them "technico-scientific" gadgets, like the "lazy spring" Slinky toy that could go downstairs, balls that changed colour when held in the hand, etc., which the family would try out together.

He had more time to spend with them in the summer vacations. The children would spend July at the family chalet in Orcières, in the Alps, alone with their grandmother Yvonne de Gennes, still strict in her principles. *"There were routines we had to keep. For example, we had to have an afternoon rest, whether we wanted to or not"*, explains Dominique. *"We had lessons in general knowledge. We also spent hours doing dictation!"*, recalls Marie-Christine. But the rest of the time, they were free to have fun and roam the countryside. Then their parents would join them and the whole family would go hiking in the mountains. *"When we were hiking, it was like having a walking encyclopaedia with us. Whatever we found — a piece of rock, a plant — resulted in explanations"*, remembers Dominique. At the end of the day, they would camp near a lake. *"As soon as it started to get cold — and at 2500 metres, that happens quickly — we would all get into the tent. Then I would tell stories from movies or books I liked. We had some wonderful days and evenings"*, reminisced Pierre-Gilles de Gennes.

Extreme Kayaker

During the vacations, the family would also explore France. *"We always had two or three kayaks on the roof of the car"*, explained Pierre-Gilles de Gennes. It was a sport he had taken up in 1964, teaching himself from an instruction manual. He was a risk taker, always looking for whitewater opportunities. One day, in Burgundy, he was just about to enter a river in full flood, but Annie would not let the children follow him. *"It was a place we didn't know"*, she recalls. So he launched the kayaks with the sister of a friend with whom they were staying. Initially, the flow was fast but manageable. Suddenly, they encountered a tangle of tree trunks and branches that had formed a huge raft across the river. It was a dangerous obstacle, but Pierre-Gilles, who was still a novice, didn't realise it. His young companion was dragged under the raft. Pierre-Gilles managed to free her, but was dragged under himself. Underwater, he thought: *"I'm a good swimmer. No need to panic."* He extricated himself from his kayak and tried to surface, but at each attempt came up below the tree trunks. The water was very murky, and he couldn't see any way to get to the surface. So he dived down and swam further, but once again found himself blocked by the tree trunks. He was beginning to panic and swallow water. He was dragged along and, on his final attempt, finally reached the surface, a hair's breadth from drowning.

In the course of their travels around France, they stayed with numerous friends — Colette Miette in Montagne de Lure, part of Haute-Provence; the family of Louis Blanchard, his former French teacher in Barcelonnette, in the Queyras region; or the family of the physician Loup Verlet — staying in big houses full of friends, children and cheerful disorder. They also visited Pierre-Gilles' aunt Edmée in Saint-Jean du Gard, whom Annie had remained very fond of since the old lady had welcomed her so warmly ten years earlier. The clocks in the house seemed to have stopped in another age. A bell was rung before each meal to warn the guests to get ready for dinner. Around the table, there was a hushed atmosphere, but Pierre-Gilles was very pleased to see his aunt and her son-in-law, Yves Laporte, who would become a professor of neurophysiology and then director of the Collège de France.

Some summers, if Pierre-Gilles had been invited to conferences or summer schools, the family would stay in the United States, in Mexico, in Canada or in Yugoslavia... For example, they rented chalets deep in the forest near the American Great Lakes, where Christian enjoyed the fishing. *"Christian learned to fish for salmon in Vancouver Bay with a physicist colleague. He liked fishing, but was less keen on the idea of us eating his catch. Sometimes, when he had caught a fine trout, he would call us over to look at it, then throw it back in the water before we got there"*, smiled Pierre-Gilles de Gennes. They also travelled several times to the Dalmatian Coast in Yugoslavia, to stay with the physicist Konstantinovitch, nicknamed Chichko. *"We were lucky enough to travel a great deal and all those trips have left wonderful memories"*, recalls Marie-Christine.

End of the Golden Age of Superconductivity

After 1965, Pierre-Gilles' interest in superconductivity began to wane. He thought it had nothing more of interest to offer. *"If you show a theorist an alloy, he can predict whether it is an insulator, a semiconductor or a metal. However, it is much harder to predict whether it is a superconductor. That is where the story of the Orsay superconductor group ended for Pierre-Gilles de Gennes, because we can't predict superconductivity or devise new superconductors, unlike semiconductors, which are so well understood that they can be engineered layer by layer"*, explains Guy Deutscher. In addition, the maximum ceiling for superconducting transition temperatures seemed to be around 25 K ($-248°C$). The golden age of superconductivity was coming to an end... Pierre-Gilles, one of its pioneers, was thinking of moving on, having published some thirty articles on the subject and a book, *Superconductivity of Metals and Alloys*, in 1964, which would become a worldwide classic, adding — as would all his subsequent works — to his reputation. Pierre-Gilles began to look for a new topic. For him, the superconductor adventure had run its course.

The only person he told of his change of direction was Jacques Friedel, presenting his colleagues, experimentalists and theorists alike, with a *fait accompli*. As head of the lab, he was concerned about who could pick up the baton and supervise the ongoing theses on superconductors. He tried to persuade Christiane Caroli, who declined and joined Philippe Nozières at Jussieu, which annoyed Pierre-Gilles de Gennes. *"If you want to waste your time in obscure problems without solutions, go ahead"*, he snapped, somewhat unfairly. In the end, Jean-Paul Burger and Étienne Guyon would pick up the reins of the experimental work on superconductors at Orsay, but only for a time. Jacques Friedel was worried about the wider consequences of Pierre-Gilles de Gennes' defection. When the captain leaves the ship, it lowers the sails... *"Pierre-Gilles de Gennes already had a great impact even then. At a time when we were well placed in the field, not only at Orsay, but also in Grenoble, and France could have had a small industrial role in this domain, his departure truncated our prospects. When high-temperature superconductors emerged 20 years later, France had to start again from scratch"*, he regrets.

Dithering

A million miles from these concerns, Pierre-Gilles was wondering where to cast his net, testing the water in different directions. He considered moving into elementary particle physics, but feeling that he wasn't good enough to find anything new and that there were already too many teams working on the subject, he gave up the idea. During the Christmas vacation of 1965, he happened to meet his old friend from École Normale Supérieure, Pierre Averbuch, by a ski lift in Merlette, near Orcières: *"I'm going to start work on hydrodynamics. I am looking at ways of measuring turbulence. It's a fascinating subject."* Indeed, turbulence is one of the great mysteries of physics. It is found everywhere, in mountain streams, curls of smoke and clouds of milk in tea. In each case, these systems seem disordered, but present similar, repetitive structures, called vortices. *"It is this intimate mix of order and disorder which gives it its charm and, it must be said, is one of its principal difficulties"*, commented the physicist Uriel Frisch.[11] Back in Paris, Pierre-Gilles began to study the velocity correlations in a turbulent flow, but getting nowhere, he eventually gave up (he would return to it twenty years later with the American physicist Michael Tabor, devising a new "drag reduction" model) (A7-8).

Profound Malaise

In truth, it was hardly surprising that Pierre-Gilles found it so hard to decide on a new subject: he was experiencing a *malaise* more profound than was apparent. *"There is, at present, particularly in solid-state physics, a sort of crisis (...). Back in 1961, [the English scientist Pippard] declared with sadness: "We are the classical physicists of our age; our science is a stately edifice, one of course with many gaps, but threatened by no transformation." (...) Pippard's observation remains broadly valid for crystalline systems. This is the most ordered and simplest phase; many of its mysteries have been resolved; numerous theorists and experimentalists are*

---

[11] Lecture at the Université de tous les savoirs, June 25, 2000.

working on the remaining problems. (...) Hence very tough competition and a feeling of anxiety (...). Personally, I am convinced that a more profound transformation is needed in solid-state physicists. They have refined techniques and also an acceptable theoretical arsenal. In these circumstances, they can and must change their goals", he would declare in his inaugural lecture at the Collège de France in 1971.

"Pierre-Gilles de Gennes was an eternal discontent. He asked himself all sorts of questions. He wanted to change fields entirely, move into nuclear physics, biology... As he was interested in biology, I put him in touch with Charles Sadron, head of the polymer research lab in Strasbourg, who had just set up a biophysics laboratory in Orléans", recalls Jacques Friedel. Charles Sadron, a big player in polymer science in France, was giving a seminar at the Orsay LPS which Pierre-Gilles attended. *"Listening to Sadron, I realised that with my background in statistical physics, I should have a lot to say about polymers..."*, he reported. Something had finally clicked.

In 1966, Pierre-Gilles started work on the science of polymers, very long molecules made up of a single repeating pattern (monomers), like individual pearls in a necklace. As he would do whenever he took on a new subject, he took stock of the existing research. This process took him to the polymer physicochemistry lab at ESPCI (Higher School of Industrial Physics and Chemistry), after the quantum mechanics class he was giving at the school. There, he got into conversation with a research assistant called Lucien Monnerie. *"Some people would have moved on immediately on learning that I was just an assistant, but not Pierre-Gilles de Gennes. I described the research we were doing at the lab and suggested a few articles"*, he recalls. That was the last he heard of his visitor until he received a note from him, the following October, inviting him to a class he was giving at Orsay on polymers: *"I'm not sure that it will interest you. It's a very modest start."* Lucien Monnerie made the trip and describes how, for his "modest start", *"he gave a superb overview of the problems raised by polymers, in other words, in six months, he had assimilated and explored all the theoretical questions in the field"*. Pierre-Gilles de Gennes also contacted polymer specialists at the Weizmann Institute in Israel and the Macromolecule Research Centre in Strasbourg, created by Charles Sadron. *"I had invited Henri Benoît, one of my distant cousins, an associate and successor of Charles Sadron in Strasbourg, to my house for a meal to talk with Pierre-Gilles de Gennes"*, recalls Jacques Friedel. Pierre-Gilles described his plan: to study polymers using neutron scattering (making use of his PhD work). He had even made an agreement with a researcher at CEA Saclay, Gérard Jannink, a former schoolmate, who was interested in the idea. Henri Benoît immediately grasped the potential of neutron scattering and was delighted to work with a theorist of Pierre-Gilles de Gennes' stature.

Noodle Soup

Once he had completed this survey of the field, Pierre-Gilles was ready. He was something of a pioneer, since he was quitting solid-state physics for a relatively unfashionable domain. Scientists interested in the subject, like the American physical chemist Paul Flory, or the English physicist Sam Edwards, were few and far between. Moreover, it was a move away from well-aligned atoms like those of metals, to long, tangled molecules resembling noodles, either in soup form (dilute polymer solutions) or combined in a dish (molten polymers). After taking on a doctoral student, Élisabeth Dubois-Violette, he began studying the statistics of simple polymers, with the hope of extending his research to biological polymers, such as proteins or the DNA double helix.

Initially, Pierre-Gilles tried to model the dynamics of a single polymer in solution, describing it entity by entity (individual monomers), as if describing the wagons in a long train. He would consider one monomer, then the next, etc., eventually reconstructing the entire polymer. Since these monomers tend to repel each other (they are said to be self-avoiding), he treated their interactions as if they were small springs. *"In 1967, we published our first article on polymers, which described a single chain that moves in a solvent and forms a coil. It was quite a significant piece of work, because it marked the start of the introduction of scaling laws into polymers"*, commented Pierre-Gilles de Gennes, who would subsequently make frequent use of scaling laws. With these laws, the variation in a phenomenon is described by means of the factors on which it depends (by a power law in the form $y = ax^b$), for example generalising laws that apply on a small scale to a larger scale, reasoning from small to large.

He followed this by describing several polymers in solution, where a chain can block repulsions between monomers. He resolved this problem using "mean field approximation" (A7-9), so that his model accurately reflected the behaviour of the polymers, but only in a very dilute regime, i.e. with very few polymers in the solution, otherwise the problem became so complex that he could not derive the simple laws he wanted. He became moody and the students felt the impact. *"What? You don't know that? I knew it 10 years ago!"*, they heard over and over again. He would even turn up in their office on a Monday morning, complaining: *"You haven't made any progress this weekend. I can see you're not interested!"* The students would keep their heads down. Fortunately, he soon decided to set aside his polymer research.

May 1968

One fine morning in May 1968, Pierre-Gilles was working in his office, windows wide open, when a flustered Jacques Friedel burst into the room: *"I wasn't able to give my class this morning. A lecturer came into the hall to urge the students to go on strike. I told him that he could at least have warned me. In any case, I have announced that I want to talk to the students after lunch."* At lunchtime, the director of the LPS met André Blandin, his loyal deputy, and Pierre-Gilles, in a small restaurant near Orsay Station to discuss their response. They were gloomy. They thought that they had escaped the strike that had gripped Paris, and here it was suddenly beginning at Orsay. Student meeting followed student meeting. Pierre-Gilles was present at the first of them. He was young enough to feel close to the students and to understand their demands — he even agreed with some of them on university reform — but his status as a teacher inevitably put him on the other side, in the firing line. In any case, he quickly lost patience with all the *"sterile verbal diarrhoea"*, as he described it, as he had at the École normale supérieure. *"There was, in the end, a lot of foolish babble"*, he maintained.

Things began to heat up: teachers were heckled and booed. Raimond Castaing, one of the founders of the LPS, was deeply upset. *"They want to bring everything down, even the work! Where will it all end?"* Jacques Friedel too was worried, but stood firm. Meeting his students, he told them that the exams would take place in June as usual. Pierre-Gilles also found himself under attack at times: *"You see, when an experimentalist gets an interesting result, he brings it to you or Friedel, like a burnt offering. Well, there'll be no more of that"*, proclaimed one student. The young lecturer went on giving his classes and the students continued to come, leaving the barricades and the meetings for a few hours, until one of them, by way of a joke, placed a radio mike on the desk before a class, with the words *"Voice of America"* on it, a reference to Pierre-Gilles de Gennes' frequent tributes to the dynamism of US campuses. When he arrived, he looked at the microphone, and without a word turned on his heels and left the hall. He would not return to the campus until the strikes were over.

He did, however, perform one "revolutionary act". He resigned from his position as a non-tenured teacher at ESPCI. *"I found that the school was run like a grocery, with grants being handed out as personal favours, etc. And the teaching was stuck in the past as well"*, he explained. He and Daniel Saint-James, also a teacher at the school, decided to pin their resignation letters to the walls of the school. The management would subsequently accuse them of publicising a "private" letter.

In May that year, Pierre-Gilles travelled to England to receive the Franco-British Holweck Prize for physics, two years after Raimond Castaing. *"The award ceremony took place at the Savoy, one of London's most prestigious hotels. I had to buy a dinner jacket for the occasion. All the English were in ceremonial dress, a myriad of decorations proudly pinned to their chests. It struck me, because it was May 1968. In the UK, old-time customs and mores were completely unchanged, whilst in France, everything was being questioned: the contrast was striking... and amusing"*, he recalled.

The revolution came to an end, but left its mark. The atmosphere at the LPS had deteriorated, so sharp was the division between "conservatives" and "leftists", which included the lab's young theoretical avant-garde. *"It made some relationships less easy than before for Pierre-Gilles de Gennes"*, notes Jacques Friedel. When, in 1969, André Guinier, Jacques Friedel and Pierre-Gilles de Gennes won an Academy of Sciences physics prize, they decided, in the post-1968 atmosphere of the time, to use the substantial sum awarded to build a cafeteria on the roof of the new Building 510 into which they had just moved, but their gesture failed to defuse the tensions.

Pierre-Gilles extended his research to polymers of a somewhat special kind, RNA (ribonucleic acid) molecules. It was his first venture into biophysics. He showed that, because some of these RNA molecules form hairpin-like loops, they can lock into each other. *"This was the cause of one of my tantrums in 2003. I was listening to a young École Normale graduate speaking about similar problems, but he was unaware of my 1968 reference"*, he commented. However, there were no experiments to back his ideas on biopolymers and they remained, like those on polymers, in suspense. At this point, he gave up work on polymers, biological or otherwise, to embark on a new adventure: the field of liquid crystals.

Chapter 8

The Liquid Crystals Adventure

"The story of liquid crystals is an epic", Pierre-Gilles de Gennes commented, paying tribute to the pioneers of this scientific adventure, which began in the late 19th century. It continues today, with relentless technological competition to improve the liquid-crystal displays (LCDs) essential to our computers, our televisions, our mobile phones, etc. What the French scientist did not say is that he made a big contribution to this epic by developing whole dimensions of liquid crystals science. Yet he entered the field almost by accident, shortly after the development of the first prototype liquid-crystal display by the firm *RCA* (*Radio Corporation of America*) in 1968. It all began when he was hailed on his way out of a class he was giving at the École normale supérieure, by his former teacher Jean Brossel:[1]

— *"One of my students, Georges Durand, has just got back from the US. Would you have room for him at Orsay?"*
— *"What was he doing there?"*, asked Pierre-Gilles.

---

[1] Jean Brossel was the co-director of the radio spectroscopy lab at the ENS with Alfred Kastler, who was awarded the Nobel prize for physics in 1966.

— *"He was doing optical measurements on liquid crystals,"* [2] replied Jean Brossel, and went on to praise the qualities of this young scientist, who had joined the army before going into research. Pierre-Gilles began to listen more attentively, wondering if there was not a potential opportunity here...

He quickly read up on the subject, in particular a summary published by the Russian physicist I. G. Chistyakov in *Soviet Physics Uspekhi* in 1967. *"I realised how poorly understood liquid crystals still were..."*, he explained. Indeed, they are strange objects: when heated, they melt, turning into a milky fluid, but instead of the disorder found in any normal liquid, their molecules adopt a degree of order, for example all facing in the same direction, like matches in a matchbox. When liquid crystals are heated to a higher temperature, they become transparent and lose all molecular order as in a "true" liquid. It is as if these objects had two "melting points", the phase between the two being called "liquid crystal", because it is both organised (as in a crystal) and liquid.

These strange liquid crystals are made up of molecules of a particular shape, for example rods or disks, which is responsible for this dual phase transition between order (in the solid state) and disorder (liquid state) (their chemical composition is also relevant in some cases). It is a little as if, because of their "awkward" shape, these molecules could not move freely during the melting process, but were forced to remain in a certain, somewhat rough arrangement, like skiers crammed into a cable car having to keep their skis vertical so as not to hurt their neighbours. This "disordered order" could be nematic (when the molecules are arranged in parallel), smectic (when, as well as being aligned, they form superimposed layers), or cholesteric (when they are arranged in a helix). These three categories are divided into numerous subcategories (and sub-subcategories), established in the 1920s by Georges Friedel

---

[2] Georges Durand spent a year at Harvard University in the laboratory of Nicolaas Bloembergen, a specialist in nonlinear optics and Nobel prizewinner in 1981.

(grandfather of Jacques Friedel), so that in all some fifteen different phases had so far been identified.

Manna from heaven for Pierre-Gilles, who had become a specialist in phase transitions since his research on magnetic materials and superconductors! *"It so happens that you can resolve certain problems in liquid crystals using the same techniques employed with superconductors, for example descriptions of order parameters. This means that Pierre-Gilles de Gennes was able to apply the approach he had developed in superconductors directly to liquid crystals, which was his real innovation. But I also think that if he hadn't done the preliminary work on polymers, he wouldn't have made the connection"*, explains Jacques Friedel. Indeed, once again Pierre-Gilles had gone off the beaten track by choosing a subject that was not solid-state physics and was not fashionable, since the science of liquid crystals had gone out of vogue since Georges Friedel's work in the early 20th century.

End of the Swarms

With this new project in mind, the theorist was not going to let an experimentalist with experience in liquid crystals slip through his fingers, so he asked Georges Durand to join him at Orsay, warning him: *"We are short of space and have few resources."* However, the young man was only too happy to be able to continue the research he had begun in the US, and seized the opportunity without hesitation. In September 1968, he started the first experimental liquid crystals team at the LPS, alongside a young researcher from Grenoble, Madeleine Veyssié, and then two postgraduate students, Liliane Léger and Francis Rondelez.

For his part, Pierre-Gilles had begun to explore one of the riddles of liquid crystals, which is why a nematic liquid crystal is "milky" (between its two "melting points"). So far, this question of the appearance of liquid crystals had not been explained, and the cause of the milkiness was still disputed. Some physicists believed that a nematic liquid crystal is milky because it is made up of different "swarms" of parallel molecules, and therefore forms a discontinuous medium. Others, however, thought that it is a continuous medium in which the orientation of the molecules varies from one to the next. Pierre-Gilles showed that this second hypothesis was correct. Liquid crystal is a continuous medium, but heat causes the orientation of the aligned molecules to fluctuate, albeit by a tiny amount, and interact in such a way that the movement of one molecule disturbs its neighbours, and so on. This cascade of fluctuations ultimately results in significant light scattering, which makes the liquid crystal cloudy. In just a few months, the theorist had ended years of argument.

"Pierre-Gilles de Gennes had done half the work to explain that there were no swarms, but fluctuations in the orientation of the molecules, when he handed over his manuscript and asked me to finish it. I sweated blood, because I wasn't really a theorist, but I got there in the end. There was a conference in the USA, which Pierre-Gilles de Gennes didn't want to attend because of the Vietnam War. So I went in his place, and presented the research to some of the biggest names in liquid crystals: "Pierre-Gilles de Gennes calculated this, deduced that, etc." At the end, they said: "You know, your student forgot a coupling there!", relates Georges Durand. Obviously, the name Pierre-Gilles de Gennes (as yet) meant nothing to them! *"When I got back, I passed on their comment. He immediately saw what it was about and corrected his mistake: he could write faster than his own shadow."* Since the young physicist thought that he had not contributed enough to the calculations, he refused to be the co-author, so this first article was eventually signed "Liquid crystals research group (Orsay)".

"We then buckled down to the experiments to confirm the model, using a light scattering method that was completely new in France and very tricky to implement", explains Madeleine Veyssié. The experimentalists had taken on a tough challenge, but it proved worth the trouble. With the very first measurements, the results exactly matched predictions. The excitement ran through the team and quickly spread beyond the walls of their lab.

Indeed, over the months, other physicists at the LPS caught wind of these successes and joined liquid crystal research at Orsay, people like the specialist on defects Maurice Kléman, Étienne Guyon — who abandoned research on superconductors — and others still, specialists in optics, crystallography, etc. All of them came together under Pierre-Gilles, who had enough ideas for everyone to get their teeth into. *"It's one of the things I like to crow about! The physicists were ready to work together on liquid crystals, which gave us a fantastic strike force"*, he admitted. For even greater efficacy, he brought three chemists into the lab to synthesise liquid crystals suitable for the experiments and to invent new ones.[3] *"Until then, we had obtained liquid crystals synthesised by chemists at the Collège de France and Thomson-CSF. And the very first samples we worked on I brought back from my travels in the US"*, recalls Georges Durand.

The Orsay liquid crystals team was becoming quite large, but this did not mean that Pierre-Gilles felt the need to organise the research, allocating specific subjects to specific researchers. He distributed his theoretical models around the lab and took no further part... as if relying on competition to stimulate his colleagues.

Creative Anarchy

On the theory side, he was surrounded by young researchers, Élisabeth Dubois-Violette, Olivier Parodi, Maurice Papoular, etc., and postgraduates like Françoise Brochard-Wyart and Albert Rapini. Although he gave them ideas, he usually then left them to work on their own, so that they would help each other to decipher the precious clues he had let slip or the equations he wrote on the blackboard (never erased after his departure!). Finally, the young theorists published articles under their own names, thanking *"Professor P.-G. de Gennes, for having suggested this research and for numerous discussions"*.

On the experimental side, tensions began to emerge: Georges Durand aroused jealousy by winning juicy contracts; Étienne Guyon would have liked to orchestrate the research, but everyone went their own way; one physicist accused another of encroaching on his territory... One team leader even went as far as to ban his postgrads from communicating with anyone on the other teams! However,

---

[3] Leszek Strzelecki, Lionel Liébert and, later, Patrick Keller.

these conflicts were short lived: the researchers knew they had every-thing to gain in sharing their results. *"Rather than a group, it was a constellation of small teams, more or less in competition. We all wanted to show what we were capable of"*, explains Madeleine Veyssié.

Even when tensions were at their height, Pierre-Gilles remained above the fray, seemingly unaware of all these conflicts, embryonic or actual. When a physicist complained to him that a colleague had stolen one of his ideas, his reply was blunt: *"I don't want to know. I don't care as long as the science progresses."* And it was progressing... by leaps and bounds! Despite, or because of, the free for all, the lab was running at full speed and generating results by the score. *"We lived through a period of extraordinary excitement, especially as liquid crystals are magnificent, their colours, the way they move... We would spend hours look-ing at them", recalls* Madeleine Veyssié.

Teeming Results

Pierre-Gilles obtained another great result on cholesterics, liquid crystals in which the molecules are arranged in a helix, like the steps of a spiral staircase. He calculated that a magnetic field, applied per-pendicularly to the axis of the spiral, has the effect of "unwinding" the structure, like bending a spring in half. He predicted that the effect of the magnetic field is to increase the height step in the helix. He began by sending his calculation to Montpellier University, out of respect for Pierre Châtelain, Montpellier's pioneer in liquid crys-tal research,[4] offering his lab an experiment on a plate. However, the weeks went by with no news. So he decided to present the result at the class he was giving on liquid crystals at Orsay (a second-year class for research students). Georges Durand and Madeleine Veyssié were in the auditorium, and immediately ran the experiment and found the predicted effect. The theorist never worked with the Montpellier lab again.

---

[4] In 1943, Pierre Châtelain showed that rubbing the glass surfaces on which the liquid crystal is placed influences the alignment of the molecules. They line up in the same direction, thereby producing a liquid "monocrystal".

Then Pierre-Gilles came up with an interpretation of the electrical instabilities in liquid crystals, giving a detailed explanation of the microscopic mechanisms involved in liquid crystal displays. So far, the first LCDs built in the US worked without anybody understanding why! Initially, he had shown no interest in the effect of an electrical field on liquid crystals... this time, it was "pressure" from the experimentalists that would lead him to a new model.

"All that was known about the early LCD screens was that without an electrical field, the liquid crystals were dark; with a field, they became bright, because they scattered light. Georges Durand came up with the idea of studying the mechanisms involved. So we made samples a square centimetre in area composed of a thin layer of liquid crystals inserted between two glass plates covered with conductive tin, and we began the experiments by varying the voltage applied between the two terminals of the cell", describes Francis Rondelez, then one of the team's postgrad students.[5]

When the voltage applied was small, the liquid crystal remained homogeneous and transparent: the sample stayed dark.[6] With a higher voltage, beyond a certain threshold, clear lines appeared at the edges of the dark areas, and formed cells, like those of a beehive. *"American scientists had already observed these strange shapes, but no one had explained them. Many people thought that they came from defects in the liquid crystal"*, explains Madeleine Veyssié. One day, however, when looking at these cells, Georges Durand noticed a dust slowly rising and falling between the electrodes, proof that the liquid crystal was in motion! In fact, the electrical field produces convection currents that make the liquid crystal flow in a loop, like washing seen through the front of a washing machine (A8-1). *"We imagined that these currents were generated by ions moving from one electrode to the other, disturbing the liquid-crystal molecules"*, explains Francis Rondelez. Pierre-Gilles

---

[5] Modern LCD screens work in a different way.

[6] Nematic liquid crystal appears milky (for reasons explained by Pierre-Gilles de Gennes) when present in a certain thickness, but it becomes transparent when thin. The sample therefore looks dark because of the transparency (the background is dark).

congratulated Georges Durand for his perspicacity, but nevertheless made no effort to explore the theory.

It took a second observation for him to change his mind. Still trying to understand the cause of these convection currents, the experimentalists modified the frequency of the field applied. Beyond a certain threshold frequency, the cells disappeared and were replaced by much tighter lines, reminiscent of a herringbone pattern. *"They couldn't be generated by the movement of ions, because the frequency was too high for ions to be able to follow the inversions of the current. It was the dipole moments of the liquid crystals themselves which were being displaced"*, they explained. Then, when an even higher voltage is applied, the liquid crystal becomes turbulent and heterogeneous: it scatters light strongly.

"I then went to see Pierre-Gilles de Gennes, a few days before Francis was to submit his thesis, with all these fantastic results, which clearly demonstrated the existence of different regimes at different voltages and frequencies. He couldn't help but be interested...", recalls Madeleine Veyssié. Indeed, a few days later, he presented an interpretation that explained the existence of the different regimes and the cause of the convection movements and of the shapes that arise, in each case describing the different contributing factors (ions, dipole moments, etc.), so that what happened at the microscopic level during the display process could be precisely understood.

Upside-Down Convection

It then occurred to Pierre-Gilles that a temperature field could have the same effect on liquid crystals as an electrical field. He formulated a few equations and gave them to the theorist Élisabeth Dubois-Violette and the experimentalists Étienne Guyon and his doctoral student, Pawel Pieranski. As she worked on the calculations, Élisabeth Dubois-Violette discovered a surprising phenomenon. Usually, heating a liquid, for example water in a saucepan, produces a convection flow: the hotter water at the bottom rises to the surface in the centre of the saucepan (hot water expands with temperature so its density diminishes), whereas the colder water near the surface sinks along the sides to the bottom of the recipient. According to Élisabeth Dubois-Violette's calculations, the convection flow in a liquid crystal is completely different, resembling a loop running parallel to the sides, whether it is heated from below or... from above (which is impossible with water) (A8-2). Pierre-Gilles initially rejected these bizarre results, then had to acknowledge the existence of this unusual convection... when it was demonstrated experimentally by Étienne Guyon and Pawel Pieranski.

"This led to years of research to develop a complete theory of the instabilities of heat convection in liquid crystals. But it all began with a few lines that Pierre-Gilles de Gennes had scribbled on a piece of paper", remarks Étienne Guyon. Nevertheless, the theorist did not sign any articles on the subject, giving all the credit to Élisabeth Dubois-Violette, Étienne Guyon and Pawel Pieranski.

Defects of All Kinds

Pierre-Gilles then became interested in defects in liquid crystals, for example the long, filament-like lines visible under the microscope, which appear in nematic liquid crystals ("nema" means thread in Greek), revealing the existence of singularities (or flaws) in the alignment of the molecules. The physicist cleverly made the connection between these threads and the vortex lines he had studied in superconductors, realising that there are similar singularities (points, lines or barriers) in systems as different as liquid crystals, crystalline solids and even superfluids. He then tried to classify and predict these singularities from the structure of the system. This was far from trivial, since they determine certain physical properties of the systems! *"I tried to think about the geometries of all the defects found in nature and I wrote a note on this subject, but it is rubbish: it asks a question without really providing an answer"*, he comments. *"One Sunday, I went blackberry picking with Barry Mazur, a mathematician at Harvard, in the Chevreuse Valley, along with our wives. You can talk while picking blackberries, so I mentioned all this stuff to him. Mazur replied: "You know, there is an extraordinary tool that could be useful: homotopy groups"*, he recalled. And on the spot, without a blackboard, the mathematician explained how these homotopy groups could be used to classify the properties of singularities. The following week, Pierre-Gilles investigated, but gave up. *"I dropped the whole thing"*, he confessed, *"though Barry Mazur had put me on the right track... A few years later, Gérard Toulouse and Maurice Kléman did some work that might be called historic, by constructing a general topological theory of defects using this same tool, homotopy groups. It shows that I missed a trick or two myself."*

Analogy with Superconductors

This missed opportunity in no way detracts from the theorist's many successes. The most important was inspired by Georges Durand. *"After the nematics, we did laser experiments on the "smectics A", where we were getting a diffraction ring for no reason we could see. I then constructed a model and defined a characteristic length. Proud of my little calculation, I went to show it to Pierre-Gilles de Gennes, who immediately grasped the significance of this characteristic length"*, recounts Georges Durand. In fact, this characteristic length reminded him of those found... in the Ginzburg-Landau superconductor theory.[7] Could there be an analogy between the smectics A and superconductors? Pierre-Gilles immediately began the calculations and noticed that the smectics can indeed be described by a wave function similar to that corresponding to Cooper pairs in a superconductor. He even showed that the phase transition between the nematics and the smectics A is similar to... the transition that occurs between the ordinary metal and the superconductor! He pushed the analogy even further by demonstrating the existence of types I and II in the smectics, just as there are type I and II superconductors, and by predicting new phases that would come to light years later (A8-3).

It was a masterstroke, an achievement that would be instrumental in the award of his Nobel Prize. It took a dual culture, both in liquid crystals and superconductors — something rare

---

[7] When presenting this subject in seminars in subsequent years, Pierre-Gilles de Gennes would initially speak of "Durand's characteristic length", which would eventually become "the de Gennes length".

among physicists — to be capable of establishing this astonishing and powerful correlation between objects that apparently have nothing in common, but in fact obey identical laws. This remarkable analogy revealed that universal laws apply in completely different fields of physics. Pierre-Gilles would subsequently go on to identify other such "superlaws", which apply regardless of the nature of the object.

Great Leap Forward

In less than three years, the physicist had made a crucial theoretical breakthrough, which propelled him to the forefront of liquid crystal research and helped to make it one of the most fashionable fields in science, attracting hundreds of researchers around the world. *"My role in liquid crystals was purely to clarify things; the real pioneer was Wolfgang Helfrich"*, was his view. Whatever he might say, he became the figurehead in the field. *"You should have seen him arriving at conferences surrounded by Élisabeth Dubois-Violette, Françoise Brochard-Wyart, etc. — charming, laid-back young women in their thirties: he created quite a stir! The Americans were amazed... In those days, women were few and far between in US science"*, recalls Madeleine Veyssié. *"At lectures, the words you heard most often were: 'As Pierre-Gilles de Gennes has suggested...'"*, adds Francis Rondelez. Indeed, the Orsay group were not always the first to see his results — sometimes he would even send his preprints to rival teams. He did not want to take the risk of being forestalled and, more generally, he did not like to let things drag on... *"Obviously for him there was no question of sitting on a result, even to give his group's experimentalists a month's head start! He was interested in getting the experimentalists competing, and to have as many of them as possible trying to verify his predictions, especially if it was difficult"*, explains Daniel Cribier. So the theorist published his models as he developed them, and... it was up to everyone else to be quick off the mark! Some of the experimentalists resented this individualistic approach.

The lectures were also an opportunity for Pierre-Gilles to network, by getting to know scientists who would occasionally send him preprints or ideas. One good turn deserves another. Few failed to succumb to his charisma. Wolfgang Helfrich was one who did. *"He is the real inventor of liquid crystal display systems. He was a funny guy. Initially, he was suspicious of me, but then he was suspicious of theorists in general. Over the years, when he realised that I wasn't trying to move onto his turf, he came to trust me. He had extraordinary beliefs: for example, he believed in that magician who claimed to be able to bend spoons without touching them. But he was undoubtedly the most inventive post-war mind in the field of liquid crystals"*, declared Pierre-Gilles de Gennes.

The Frenchman was equally ready to invite researchers he met at these conferences to stay at the LPS, and friendships sometimes grew out of these meetings, for example with Paul Martin, a condensed matter physicist at Harvard. *"We had epic discussions on the dynamics of liquid crystals. Initially we had very different points of view. I made a bet with him, which I won: it was for a dinner at the Tour d'argent!"*, he remembered with pleasure. Another person who became a friend was Bob Meyer, a physicist who came to LPS to do a postdoctoral course. While they were queueing for lunch at the Orsay canteen, the young post-doc was thinking aloud, playing with the idea of molecular symmetry in smectics. *"If all the molecules were chiral, wouldn't the layers produce an electrical polarisation?*, he mused.[8] Pierre-Gilles realised that what the young physicist was doing was nothing less than inventing a new liquid crystal phase. These would become the so-called ferroelectric smectics, liquid crystals with spontaneous polarisation, of particular interest for display systems, which would shortly be synthesised by the chemists in the Orsay liquid crystals group.

In fact, this was one of the great successes of the Orsay chemists, sealing their acceptance by the physicists, which was far from guaranteed. *"What is needed, if you want physicists and chemists to collaborate, is for them both to move outside their basic discipline. Both sides have to a certain tendency to corporatism. The obstacles to such collaboration are a very widespread intellectual laziness, a compartmentalisation that is already*

---

[8] A molecule is chiral when it exists in two forms that are mirror images of each other, like two hands.

apparent to students in university structures, and above all the effort that such an undertaking demands. It is a long-term process, which requires a great deal of determination", wrote Pierre-Gilles in 1973. The physicist would always seek to break down the barriers between the disciplines. In fact, encouraged by this fruitful experiment, he would later bring chemists into his lab at the Collège de France, but with less success, and there is no doubt that the multidisciplinary nature of ESPCI would affect his decision, in 1976, to become director of that school.[9]

While his desire to promote the flow of ideas prompted him to invite numerous physicists to stay at the LPS, he was also a frequent visitor to foreign laboratories, for example the IBM lab in Zurich, where he was now a consultant. *"At the time, this lab was relatively unknown on the world stage, but the research being done there was fascinating. It became famous after winning two Nobel prizes on the trot, one for scanning tunnelling microscopy in 1986 and the other for high-temperature superconductors in 1987. This prize went to an old pal of mine, Alex Müller, a clever and interesting guy whom I knew well, as he was doing magnetic resonance experiments in potassium doped graphite and, because I knew about resonance through Abragam, I helped him to model them"*, he explained. On his visits to Zurich, the French physicist stayed with his friend Peter Franck, who was responsible for scientific intelligence in Europe on behalf of *General Electric*. The two men got on like a house on fire. A bachelor, Peter Franck always had a beautiful woman on his arm, a big attraction for his friend.

Almost every year, Pierre-Gilles also visited the *General Electric* laboratories north of New York. He would take the opportunity to see Maria, his father's former governess, who now lived in Gloversville, around 60 kilometres from there. She had ended up in this small American town after Pierre-Gilles' father's apartment was sold to glove manufacturers originally from Gloversville, and she had continued to work for them, eventually ending up in the service of two *"rich and unbearable"* old ladies. Maria enjoyed Pierre-Gilles' visits

---

[9] See Chapter 12.

and would show him photos of his father, Robert de Gennes, describing the life he had led before and during the war. One year, Pierre-Gilles knocked on the door of the imposing Gloversville house and asked to see Maria, only to learn from the old ladies that she had died. Pierre-Gilles was indignant not to have been told and asked for his former nurse's belongings. But they had all been thrown away. He went away, deeply saddened.

Imperial General

At the end of 1970, the physicist was attending an international conference in West Berlin, which in his opinion marked the start of the "great leap forward" in liquid crystal physics. Despite the interesting nature of the lectures, he took a day off to make his way to East Berlin. After undergoing the usual tedious checks, he explored long streets lined with buildings as sinister as the atmosphere around them. Back at the conference, he attended the evening lakeside party. For a lark, he and a few of the other scientists dived into the lake for a swim. Suddenly, they heard a gunshot. The swimmers realised that they were probably approaching an invisible border with East Germany and swam back as fast as they could. The day did nothing to alter the Frenchman's dim view of communism.

He was beginning to wonder whether he had not gone as far as he could with liquid crystals, but said nothing about it. In addition, he was beginning to feel less comfortable at the LPS, which had become a big structure (more than 150 researchers) headed by a scientific council which was not always as flexible as he would like. But what should he do? When Anatole Abragam hinted that a professorial position was available at the Collège de France and that he would have a chance of getting the job, he saw a way out of his problems. A prestigious offer like this is not something you reject! Although he could have held the chair at the Collège while continuing his research at Orsay, he immediately chose to transfer to Paris, seizing the opportunity to create his own lab. Once his appointment at the Collège de France was confirmed, it became obvious that he would terminate his research on liquid crystals. However, unlike in the case of superconductors, the work on this topic continued at Orsay. It should be said that he did not bring his research to an abrupt end, as he had with superconductors. He continued to attend conferences on the subject over the next few years, for example at Pont-à-Mousson in 1971, an event he even helped to organise. Nevertheless, he seemed elsewhere. He doodled parallel lines of liquid crystals. *"If you look closely at this drawing, you will see that it actually, secretly represents a woman's hair"*, he confessed. He would continue to publish articles on liquid crystals until the late 1970s — even a final article in 1990 — and in 1974 published a textbook on them, *The Physics of Liquid Crystals*, which remains a standard textbook.

Missing the Industrial Boat?

Pierre-Gilles witnessed the tremendous development of flat screens in the early 1990s, responding to the indignation at the lack of French patents and production in the field, given that the country was in the forefront of research. *"It's true, we could have done better"*, he commented laconically. He would even take some of the blame: *"We weren't worried about filing patents; in hindsight, I realise we should have been a bit more serious (...). At the time, there was no problem with research funding and it didn't really occur to us to file patents..."* In fact, these criticisms were unfounded for two reasons. Firstly, his own results related to very fundamental physics, and were totally unsuited to any kind of patenting. And secondly, the French industrial community knew all about the progress in liquid crystals at the LPS, and its potential applications. *"In 1969, we presented our results to researchers at Thomson, at the instigation of Pierre-Gilles de Gennes. We thought that our results might interest them, but nothing ever came of it"*, recalls Élisabeth Dubois-Violette. Industrial firms like Thomson and Philips did do serious research, but manufacturing flat screens requires considerable investment. *"I don't think that the researchers had anything to reproach themselves with. I put a lot of effort into applied research, but after that it is the industrialists who decide..."*, explains Georges Durand.

When Pierre-Gilles left Orsay, some people accused him of caring only about his own interests and abandoning his troops. They likened him to an imperial general, who puts together an army, rampages across the battlefield wielding his sword, and then, when he has finished, begins another campaign, leaving his troops behind. All he wanted was to fight shoulder to shoulder with other scientists, but no one could follow him: he moved too fast.

Chapter 9

Out of the Tunnel

The early 1970s were a watershed for Pierre-Gilles de Gennes, both from a professional point of view — he left the Solid-State Physics Laboratory at Orsay for the Collège de France — and from a personal perspective. The successes he had stacked up since the beginning of his career had placed him amongst the great figures of contemporary science, but rather than enjoying this success, he was worried... about mortality. It was the start of a troubled period in his life.

As he did every winter, in February 1972, the scientist took a few days to go skiing in the Alps. Around 4 p.m., he would leave the slopes and shut himself away in his hotel room to read articles. One evening, a physicist friend commented: *"Don't you ever stop working?"* Pierre-Gilles smiled: *"You know, I probably won't live long. So I can't waste any time."* In response to his friend's questioning look, he added: *"My father died young, of heart disease, and his father before him. It seems to be hereditary."* It was not a joke: approaching 40, Pierre-Gilles genuinely thought he was in his final years. In fact, he did not feel in good shape. *"I had always had problems with hayfever — it's genetic, as I had a grandmother who they said couldn't travel in an open carriage because of it — and I remember how, on my honeymoon on the motorbike with Annie, the plane trees of the South made me sneeze and my eyes run. But in*

1967, I developed asthma, which quickly became so severe that I never went anywhere without my medication." Unfortunately, the bronchodilators he used were contraindicated for people with weak hearts, which increased his fear of succumbing to the disease that killed his father and grandfather. He began to complain of shortness of breath and gradually gave up his kayaking trips. He had to stay away from his much loved dogs. *"It was the first thing I had to do, according to the doctors"*, he mourned. Soon after, he developed such bad back pain that he underwent major surgery, which was a blow to his morale. Despite the instruction to stay in bed, he started working again before the convalescence period was over — he couldn't stand inactivity. *"During his two weeks in hospital, I went there almost every day to work with him, at his request, even though he was still suffering from the after-effects of surgery"*, recalls Phil Pincus.

The thought that he only had a few years left to live — a near certainty with him — became obsessive. But it wasn't so much about his health as about the passage of time. He felt himself ageing, and found it hard to deal with. He always worked so hard that he had not noticed. May 1968 opened his eyes. The students did not see him as one of them: he realised that he had crossed the Rubicon. In recent years, he had also seen several friends die young. At the end of July 1968, on vacation in the Var, he had wanted to introduce Peter Franck to kayaking. *"Come on, just a short paddle! The water isn't that cold!"*, he had teased, surprised by the American's categorical refusal. A few weeks later, his friend died during a cardiological effort test. His friend from lycée, Boris Bespaloff, had died in a car accident in 1970, and Maxime Guinnebault, his classmate from École Normale, a few years before. He had been hard-hit and left shaken each time.

His wife noticed that he was sometimes moody, but refused to take it too seriously and would give as good as she got. *"Everyone always indulged him — his mother most of all — but I never hesitated to straighten him out when necessary"*, she emphasises. All in all, though, the physicist got on with his life — however close the end might be — pursuing his research with unalloyed interest. Then came an opportunity to change everything. In his office one day, he received a call from Hendrik Casimir, head of research at *Philips: "We would like to come to see you"*, declared the Dutch physicist. *"With pleasure. Which labs would you like to visit?"*, asked Pierre-Gilles. *"We'll see, we'll see..."*, replied his caller evasively. The Frenchman met Hendrik Casimir in person off the plane at le Bourget airport, and was preparing to head for Orsay when his guest gave him the address of a top Paris hotel: *"We would rather speak to you in private"*, he explained. Once comfortably settled in a suite, he put his cards on the table: *"I'm thinking of retiring. We thought of you as my replacement. We would like you to come and head research at Philips."* Stunned, Pierre-Gilles asked for a few days to think. *"Work in industry. Go abroad. Why not?"*, he mused. However, he did not feel he had the skills for the job. And life in Eindhoven did not attract him either. In the end, he refused and let the opportunity to change his life go by. Nevertheless, change it would. *"The year 1972 was an important milestone"*, he admitted.

In August 1971 — not long before his 40th birthday — he went to a summer school in Haifa in Israel. The prospect was neither pleasant nor unpleasant. The subject was the physics of liquids and left him a lot of free time, which he spent in company with a cheerful gang of French students, amongst them Maurice Papoular, Sébastien Balibar, Sauveur Candau and Françoise Brochard-Wyart, his graduate student at the LPS, who was there with her younger sister Dominique. In late afternoon, the whole group would go to the beach. The atmosphere was cheerful and light-hearted. At the weekend, they all crammed into rental cars to visit Megiddo, a fabulous archaeological site where 20 towns had grown one on top of the other over the previous 7000 years, then Montfort, a fortress dating from the Crusades, then other sites. The physicist enjoyed himself, driving fast on the rough roads, claiming that it evened out the jolts from the potholes, and then going off into complex explanations of dubious credibility about resonance frequencies. Both Françoise and her sister were attractive blondes, who turned heads. In their short, 1970s style dresses, they were even refused entry to a monastery. *"When the old monk opened the door, he averted his eyes and cried "Vade retro satanas". I roared with laughter"*, recounted Pierre-Gilles de Gennes. During this stay, the physicist relaxed and found his eyes drawn to his attractive student.

He never disguised his interest in women — he enjoyed flirting and had occasional affairs. Annie always found out in the end, but he responded to her reproaches with Olympian calm. *"In life as in his work, he was attracted to the new"*, she observes. She chose not to attach more importance to these affairs than they deserved.

Back in France after the summer school in Haifa, Pierre-Gilles immersed himself in work again. He spent the winter vacation of 1972 in Zermatt, one of Switzerland's great resorts — *"Annie had made real progress and we had a great time skiing together"*, he reported. Then in March, he went to Harvard, happy to touch base with colleagues like Paul Martin, Michael Tinkham and George Benedek. He also visited the *Kodak* research centre in Rochester, then travelled to Providence, where he met Leon Cooper, discoverer of the eponymous electron pairs, who was now focusing on biology. *"I also saw my friend Franco Fido, who was teaching literature there."* In between travels, life seemed to go on as usual, but in fact, at the lab, Pierre-Gilles was unable to take his mind off his student, whom he saw every day (she worked in a neighbouring office). *"Before the summer, I had known her only slightly. I just knew that she was the niece of Jean Wyart* (Author's note: a French crystallographer and member of the Academie Française), *since he had asked what I thought of her when she was doing her thesis. I had answered cautiously: 'She seems bright enough. We'll have to see'"*, he smiled. He was probably not indifferent to the young student's admiration for him — it was his pygmalion side. The weeks passed. He invited her out to a restaurant, once, twice. *"We got into the habit of having lunch together and eventually ended up in each other's arms"*, he recounted. Over the months that followed, they improvised outings and took advantage of conferences to spend time together. Their affair quickly ceased to be a secret.

Annie had guessed everything weeks before her husband told her. She realised that, this time, it was more than a passing fancy, and wanted to leave. He stopped her. She asked him to leave. He refused. He ended the relationship with his student, and then began again. There followed four years of successive break-ups and reconciliations, renunciations and resumptions. He was incapable of putting an end to these unending heartbreaks, incapable of sticking to his decisions, incapable of escaping his passion. Torn between the two women he loved, he was for once unable to make a choice.

Le Boudin Sauvage

To deal with things, Annie threw herself heart and soul into a remarkable venture: opening a restaurant at home, in her own house in Orsay. She had always loved cooking. Even back in Berkeley, she had been a good cook and would compete with her friend Josie Fido over who would prepare the best dinner! Since then, she had become a genuine cordon-bleu. But was it enough to start a restaurant? Her decision caused a general outcry. *"It was terrible ! My mother said I was mad and my mother-in-law that I was dragging the de Gennes name "through the mud" — imagine: a de Gennes as a lowly cook! But I held my ground"*, she recalls. She did nothing by halves. She was eager to succeed and went all out to create new recipes. She travelled to Japan to find the right dinner service, linking a different plate with each dish. She redesigned her house from top to bottom to fit in the tables, and in 1976 finally opened *Le Boudin Sauvage.*[1]

Initially, the tables were confined to the ground floor garden room. Then, as the restaurant became successful, she added more upstairs, even... in her husband's study! In the end, five rooms in the house were allocated to the restaurant, serving up to 50 people a sitting. *"She worked hard, getting up at three or four in the morning to find the best products at Rungis"*, recalled Pierre-Gilles. On the menu were calf sweetbreads with aubergine preserve and crushed hazelnuts, perch with garlic preserve and rosemary honey, prawns sautéed with vegetables and coconut milk... She scored 17 out of 20 in the Gault et Millau restaurant guide! Here she was, on an equal footing with the top chefs. Everyone was impressed by her success, even ... Yvonne de

---

[1] It was Pierre-Gilles de Gennes who chose the name of the restaurant, an allusion to *La pensée sauvage* by Claude Lévi-Strauss.

Gennes, who liked nothing more than to converse in English with the foreign guests. *"She was delighted by the lively atmosphere in the house. She also became much closer to her grandchildren, even though their world was different from hers. She had a good end to her life: she wasn't bored, she had a beautiful garden, she was spoiled and regaled with excellent meals, etc.",* rejoiced her son. *"At the end, it is as if she realised that she had missed out on a lot of things in her life: her grandchildren, family life, feelings, etc.",* suggests Annie. *"She was probably sad that Pierre-Gilles was not warmer with her. He wasn't really an attentive son. But there is no doubt that she adored him... After her death, in 1983, for example, we found the letters that we had written her from Berkeley, but she had only kept the passages written by her son!"* Now, in charge of her restaurant, Annie no longer had time to think and found herself... all the better for it, especially as in 1976 the situation would become clearer, though in no way less unorthodox.

Second Youth

In June 1976, Pierre-Gilles was staying in Tromsø, a small town north of the Arctic Circle in Norway, after giving a series of lectures. The sun never sets at that time of the year, so the town was bathed in a strange nocturnal light that gave it a slightly unreal atmosphere, reinforced by the fact that snow was falling despite the season. It was here that Françoise told him that she was expecting a baby. He didn't hide his anxiety. The following year, in February 1977, while he was at a summer school in Peru, he received a telegram announcing the birth of his daughter, Claire.

The new baby inevitably changed many things. Pierre-Gilles now slept one night a week at Françoise's house in Saint-Cloud, then two, then three when his son Matthieu was born the next year, in May 1978. He nevertheless continued to spend a lot of time at home in Orsay. His older children were already grown up — now aged 20, 22 and 24 — but they were a close family, and saw a lot of each other: family meals were sacred. The table was set in the garden in fine weather.

With this new relationship and young children, it was as if Pierre-Gilles was enjoying a second youth. He re-experienced the joys of fatherhood, but also its worries and demands. Matthieu was born prematurely. *"He spent a month in an incubator. Over the following months, we were continually worried, because he wouldn't eat. I can still see myself in Savoie, spending hours, trying to get him to feed from the bottle"*, he recalled. Then Olivier was born in August 1984 and Marc in January 1991. *"I looked after Olivier from the moment he was born. Three weeks later, I took him hiking in the mountains — I carried him in a baby-pouch on my chest. Perhaps it was a bit unwise?"*, he wondered. To look after his youngest children, he gave up his functions as professor-at-large at Cornell University — *"I didn't want to leave the little monkeys"*, he explained — and, for example, refused the Eberly chair at Penn State University in 1988 on the grounds that he had *"too much family on this side of the Atlantic"*.

He took these decisions without the slightest regret. While at the beginning of his career, he had been exclusively preoccupied with his work, he now had nothing to prove in science. He was more available and devoted several evenings a week to his young children. *"It is as if he felt lost when the children were not there"*, explains Françoise. He helped them with their homework, which was not always easy. There was nothing he liked more than to restate the terms of a problem or, once the answer had been found, to suggest looking for other solutions, with the result that homework took much longer than expected — and the children ended up tearing their hair! Their father was absolutely insistent on them thinking. *"I am often horrified by my own children when, faced with a physics problem, they try to come up with a figure without thinking"*, he confessed in one interview.[2] He did not look at their school reports, because he wanted them to understand that coming home with good grades was not the most important thing.

The children grew up and adjusted willy-nilly to having a part-time father. Everytime he arrived, it was a celebration. They would wait for him and, as soon as they heard him come in, rush down the stairs, jostling to be the first to touch him. At weekends, they would walk together in the Bois de Boulogne or visit an exhibition, all

---

[2] *L'Actualité scientifique*, December 1993.

sitting on the floor in front of one work or another, sketching. Back home, a moment would come when the children realised that their father had left: as he did not like goodbyes, he often slipped out without saying a word. Despite his absences, they grew up with the sense of having a dependable father they could count on. *"He was the pillar who held our family together"*, declares his daughter Claire.

By the early 1980s, the situation had stabilised, but at what cost? Pierre-Gilles now divided his time between his two families, one evening at home, another at Saint-Cloud. Although his "big children", as he called them, were now adults, he split his weekends and vacations to have time with all of them, spending July with one family and August with the other. *"I used to love reading books aloud, as my mother had done, but unfortunately I did it mainly with my second set of kids. With the first, I was too busy and Annie was tremendously efficient, so I wasn't as present, whereas with the others I realised that life was going by and these children didn't have the same family status: they were less protected. So I tried to look after them as much as possible."*

Pierre-Gilles cared little for the opinions of others and did not do much to cover up this double life. He would would either mention it casually to his friends, or present them with a *fait accompli*, for example turning up with Françoise on his arm. He would also sometimes attend official ceremonies with her. At a sumptuous dinner given at the French Embassy in Lisbon in 1979, the French physicist was in the place of honour opposite the ambassador, whilst Françoise, who had no official status, was relegated to the end of the table, opposite a scientific attaché. Ultimately, some of the people around him knew what was going on, others did not. Logistical headaches, complicated diaries: there was finally some order in the chaos...

Despite this complex private life, the demons that had assailed him with the approach of his 40th birthday had retreated. He retained his solid base at Orsay, while his young children gave him the illusion of a second youth. He got out his pencils and returned to the Louvre to sketch his favourite works, *The Winged Victory of Samothrace* and *La Belle Allemande* or Gregor Erhart's *Mary Magdalen*. He took up sport again, learning to windsurf on Lake Annecy, and practising when he could, with varying degrees of success! In 1981, at a conference on random media at Carry-le-Rouet, near Marseille, he rented a windsurfer between two lectures. Smilingly, he recalled how he had fallen on some sharp rocks and return painfully to the conference, grazed and covered in blood, quickly recovering his smile and his self-confidence following a shower and a change of clothes! He was no longer fragile as he had been in his late 30s, when he seemed anxious about wasting time, about wasting his life, looking for some ill-defined renewal... This dark period was over. He had captured the rose, along with its thorns, and no one would hear him complain.

Chapter 10

Recognition

"This is no way to treat a professor of the Collège de France!", protested Charles Ryter, an astrophysicist at CEA Saclay, paddling hell for leather behind Pierre-Gilles de Gennes. The two scientists were kayaking down a river in the Doubs region, when they paddled straight into the middle of a fishing competition. Pursued by volleys of curses, they fled as fast as possible, fishhooks whistling over their heads. Once safe, they burst out laughing, Pierre-Gilles teasing his companion for wanting to defend his professorial honour.

This Chair was the greatest distinction in higher education, the Collège de France having been a centre of excellence in French research since its creation in... 1530. It had maintained its high level as a result of a rule stipulating that, when a chair became vacant, the professors must elect a new holder who would *"represent the latest developments in science and be responsible for setting the next research topic"*. This meant that the research done at the Collège had kept pace with scientific progress. So having submitted his application (at the suggestion of Anatole Abragam, himself the holder of a chair), in November 1970 Pierre-Gilles undertook, as required, a sort of electoral campaign amongst the professors of the Collège. *"I didn't manage to see all 52 — it was too complicated — but I met many from different horizons: Egyptologists, biologists, etc. And I had the great pleasure of conversing with Raymond Aron, a man I greatly admire. I met him once or twice again after that, but I'm sorry I didn't know him better*", he regretted. Pierre-Gilles was easily elected in September 1971, succeeding the physicist Jean Laval.

The new "Chair of Condensed Matter Physics" was far from purely honorary. Every year, the professor was obliged to give a new class on a subject at the cutting edge of research. On the other hand, he was free to do his research at the Collège or elsewhere. Pierre-Gilles could therefore have continued to work at Orsay, but his dream was to create a small, effective and adaptable experimental structure to underpin his theoretical research. In addition, his goal was to position his laboratory at the interface between physics, chemistry and biology. The fruitful collaboration between physicists and chemists at Orsay had reinforced his interest in working directly with chemists, so why not with biologists? The Collège, the epitome of multidisciplinarity, could be the right place. At the time, crossdisciplinary research was far from a commonplace idea, and Pierre-Gilles was something of a pioneer. Above all, however, he had set himself an ambitious goal in creating his own lab.

In May 1971, he visited Étienne Wolff, director of the Collège, to discuss the conditions of his entry. *"We have allocated you premises on the first and fifth floors. For the moment, they are still occupied by your predecessor's research team, at least thirty people. Push them out or put the pressure on, but they have to go!"*, declared the director. Seeing the new professor's stunned expression, he added: *"The Collège does not exist to maintain moribund teams. It's not a charity! Renewal is vital."*

Pierre-Gilles agreed in principle, but was distressed by the idea of having to fire thirty people. The interview hit him like a cold shower.

He made an appointment with the old team and set off nervously to his future premises. The Collège's physics and chemistry building was as shabby as its main courtyard was majestic. Pierre-Gilles passed through the mosaic-decorated hall, framed by a huge staircase that spiralled around an old wire elevator cage, but chose to take the stairs, four at a time. He opened the door of the first-floor laboratory and was dismayed by what he saw. The half-empty room was dismal. The walls were painted dull beige and khaki. He moved on quickly, and went to meet the researchers. *"Fortunately, the problem was resolved, because many of them had already found positions elsewhere"*, he was relieved to find. The remaining few accepted the jobs he offered them in the future structure.[1] As he left, he felt a weight come off his shoulders, even though the lab would have to be built from scratch.

First Course at the Collège

For the moment, he wanted to concentrate on the class he would have to give at the start of the academic year. *"It's not a soft option teaching at the Collège de France! You have to develop a new course every year, whereas usually as teachers, we more or less repeat the same class several years running, with minor adjustments. In addition, the level is very high, with an extremely demanding audience. Pierre-Gilles de Gennes thought that some of his listeners would be better than him, for example in maths. So he would disappear for weeks at a time to prepare his classes at the Collège"*, recounts his physicist friend Julien Bok. It is true that, unlike other professors there, the theorist always obeyed the requirement to begin a new topic every year (though he dropped the other classes he had been giving). From then on, these preparations would take so much time that he would usually go away on summer vacation with a suitcase full of articles. Indeed, this period was not only enriching — he would browse through and assimilate dozens of articles, a remarkably effective form of "continuous education" — but also productive, because the reading would help him think about outstanding problems and often lead to publications. All in all, by the end of the summer, he would have mastered a new field, and put across the essence of it in his classes. *"I am sure that this*

---

[1] Jacqueline Jouffroy became administrative director of the laboratory. Jean Billard and Paul Levinson pursued their research. The two postgraduates, Hubert Hervet and Raymond Ober, agreed to spend two years learning about neutron scattering and then to come to work at the lab.

requirement teaching was very constructive and gave him an overview of different fields", comments Jacques Friedel.

"He did the work of clearing the undergrowth and doing the heavy chewing: all we had to do was swallow the summaries he gave us in his class", adds Françoise Brochard-Wyart. Aware that they were getting a bargain, people came in large numbers to listen to him, sometimes travelling a long way. *"Every week we would see the condensed matter physicists "migrating" towards the Collège"*, one physicist jokes. Experimentalists and theorists alike came to the lectures, because even though the calculations sometimes went over their heads, the experimentalists found that the classes not only broadened their horizons but also gave them ideas. *"On one occasion, Pierre-Gilles de Gennes mentioned a phenomenon called "roughening transition". There was a certain factor that linked the existence of facets on crystals with the latent melting temperature* (Author's Note: heat required for a phase transition). *As I listened, I said to myself: "Usually, you have to heat a crystal to make it melt, but for helium crystals, in certain conditions, it's the opposite: they crystallise when heated! Now, in helium, the latent heat changes sign..." I realised that there was something worth exploring here and it started me on ten years of research, in collaboration with Philippe Nozières"*, recounts the physicist Sébastien Balibar. In this way, Pierre-Gilles de Gennes, known as "PGG" in the physics community, inspired a multitude of research topics and experiments, sometimes without realising it. Conversely, when he explicitly suggested ideas for experiments which were not taken up — though it was rare — he would not hide his irritation, commenting to no one in particular: *"It took the experimentalists months to grasp the importance of this idea and to decide to work on it."*

His classes were like nobody else's. He would mount the podium with obvious relish, like an actor on the stage, rehearsing his effects and tracking his audience's reactions. He would captivate them like a snake charmer, so that in his mouth the most complex theories seemed as clear, straightforward and limpid as mountain streams. However, his classes, often structured around a handful of images, were simple only in appearance. Their content was both technical and complex. Some enjoyed them anyway, like a landscape seen from a train, without the details. But this was not the case for everyone.

"He puts on his show. He doesn't care whether people understand or not ", one physicist complains. In fact, while the auditorium might have been crammed to the rafters at the beginning of the year, full of admiring faces and lyrical comments, gaps in the ranks would soon appear at subsequent classes, as the nonspecialists swiftly dropped out. The fact is that the course was aimed at a particular coterie of researchers, for whom Pierre-Gilles had such authority that his words became, despite himself, gospel truth. *"His class resembled a high mass, where some people had no greater ambition than to be cited by the "master", as if receiving a papal blessing. One word from him was the ultimate consecration! "*, frowns another physicist.

Dusting Down the Lab

At the same time, Pierre-Gilles had to set up his lab, starting from scratch, or rather... from rooms stuffed with used files and old equipment. *"Fortunately, I managed to bring in two top physicists, Madeleine Veyssié, a team leader in liquid crystals, and Christiane Taupin, a highly competent experimentalist "*, he recounted.[2] Their first task was to equip the place with offices and experimental labs. Although he had warned them, the two women were horrified when they saw the dark and shabby premises. However, they summoned up their courage and spent several weeks at the end of 1971 sorting and clearing. Fortunately, they made some good finds in the bric-a-brac, for example unearthing in a dusty corner the original desk used by Léon Brillouin,[3] a great French physicist who had held a chair at the Collège in 1932. Gradually, the lab took shape. *"All the money went into buying equipment: we began by acquiring an EPR system* (Author's note: electron paramagnetic resonance, a technique used to identify and locate magnetic species), *then an electromagnet. We didn't spend a single centime on paint or furniture "*, reports Madeleine Veyssié. They

---

[2] Christiane Taupin was a student of André Guinier. After her thesis, she joined Pierre-Gilles de Gennes, who sent her to train at Harden McConnell's laboratory at Stanford on the physical chemistry of biological membranes.
[3] See Rémi Mosseri, *Léon Brillouin. À la croisée des ondes*, Belin, 1999.

also set up a chemistry lab to Pierre-Gilles' specifications, which would soon be occupied by Maya Dvolaitzky, from the Collège's chemistry lab (headed by Jean Jacques), swiftly joined by another chemist, Marie-Alice Guedeau-Boudeville, or "Booboo". Since Pierre-Gilles never called her anything but Booboo, foreign visitors would regularly turn up at the lab asking for "Madame Booboo"!

Early on, Pierre-Gilles had chosen the biggest room on the first floor for his office, where he installed two armchairs that gave the place the atmosphere of a British gentleman's club. On his library shelves were various legacies to the Collège such as the original works of Laplace or a picture of Jean Perrin, along with photos of his children and personal mementos. Then, after completing their doctorates at Orsay, Liliane Léger and Francis Rondelez joined the team — the fifth floor was arranged for the purpose — and the research force soon grew to some thirty people: the ideas sown by Pierre-Gilles gradually germinated, and the laboratory took root.

Birth of Soft Matter Physics

When he arrived at the Collège, he had only a vague idea of the experiments he wanted his two colleagues to do. He suggested research on bidimensional systems and hydrodynamics. He asked them to begin studying monolayers and membranes, i.e. amphiphilic substances such as soap (also called tensioactive agents or surfactants). These are molecules that have a water-loving (hydrophilic) head and a water-fearing (hydrophobic) tail. When immersed in water, they unite to keep their tails out of the water. For example, they form surface films (heads in the water and tails in the air), or micelles, minute spheres with in-turned hydrophobic tails, bilayers (superimposed films), tubes or sheets, and so on. All these objects had been known for a long time, but Pierre-Gilles wanted to study their two-dimensional phase transitions, for example, the effects of concentration or compression. Then he would move on to interface films, for example explaining how soap bubbles form and die; why minute particles (colloids) remain suspended in a fluid or conversely merge and sink; what happens in emulsions (where two liquids, like oil and water, do not mix but form drops one inside the other). From interfaces to surfaces, it was only a step to wetting, explaining the spread of a liquid across a surface, and from there to adhesion, which combined what he knew about polymers — he would go back to his research on "noodles" to make a decisive break-through — surfaces and wetting.[4]

All these everyday objects (bubbles, emulsions, glues, etc.) were already familiar to chemists or engineers, but had so far been ignored by physicists. Using the tools of statistical physics, in partic-ular the renormalisation group, discovered in 1971, Pierre-Gilles would transform them into objects of science and build a new scientific field: soft matter physics, a term invented by Madeleine Veyssié for the liquid crystals she had studied at Orsay. This disci-pline explores objects that are neither entirely solid nor entirely liquid, characterised by large reactions to small actions. It explains the properties of materials by considering objects not at the scale of electrons and atoms, but at the scale just above. *"At the time I started working, the whole focus was on description at the atomic and quantum levels. Over the years, however, I came to feel the importance and universality of higher scale models, of the kind you like"*, he explains in a letter to Louis Néel in 1978. *"For example, our approach to liquid crystals bears the stamp of the magnetism you created! These days, I'm slightly concerned to see so many youngsters (...) ignoring problems in macroscopic physics (for exam-ple in rheology)."*

Under the Dome

His appointment to the Collège de France reflected the reputation Pierre-Gilles had acquired, but it was the litany of results on soft matter that consolidated it, bringing showers of distinctions from the 1980s onward. For example, the French Academy of Sciences opened its doors to him in 1979. Apart from the prestige, the insti-tution performed different roles, for example it maintained *"vigilant oversight on the place of French research in the world, on the organisation of*

---

[4] The details of this research will be described in the next chapter.

research, on the direction of scientific programmes, and on (...) the applications of the sciences." Realising that the institution in fact had no power, the physicist soon came close to feeling that the *"old and respectable firm"* was completely pointless, though initially he was full of goodwill. Thus, when the Academicians met to decide whether to allow the publication of articles in English in the *Proceedings of the Academy of Sciences* (*CRAS*) — a publication of which he was fond — and to open their columns to foreign scientists, he attended the debate, surprised that the question aroused so much discussion, so obvious was it to him that the *CRAS* (and the influence of French research) had everything to gain from such a measure. After half an hour, the Academicians reached agreement, but one of them then reopened the discussions by asking: *"Ultimately, what does it mean to be foreign?"*... Pierre-Gilles emerged from this session exasperated, determined not to go back. *"It's typical of the Academy... Between pointless debates and the power of the pressure groups, there's not much one can do...",* he regretted. Indeed, he also — and above all — objected to the existence of "chapels" within the institution. *"a) The Paris region is overrepresented; b) Appointments are controlled by a few powerful groups (e.g. atomic physics, statistical physics). Researchers working on new fields find it hard to gain acceptance",* he explains in a 1999 letter. So he attented less and less often, and not at all after 2000. *"I have little influence at the Institute, I hardly have the time to go there and I'm not greatly convinced by certain decisions. (...) As regards the elections, there is also a fiddle going on at the moment, with the Academy converting its associates to members, rather than recruiting new blood. I stood down from several committees because of this",* he replied in March 2002 to a researcher who reproached him for what he saw as the unreasonable absence of a great glass specialist from the Academy of Sciences.

His defection would attract harsh criticism from certain physicists. *"He never sets foot in the place. He criticises but never lifts a finger to change anything",* they complained behind his back. He responded by saying that he felt that any attempt to change the Academy was futile. *"Everybody who tried had gotten nowhere",* he asserted. When he became president of the *Physical Chemistry Society* with the remit of bringing some dynamism to this small structure and its journal, all his suggestions fell on deaf ears. "*I have bad memories of the* Journal français de chimie physique. *I tried to reform it a long time ago (when I was president of the Society), but I encountered huge inertia*", he told someone who had written requesting an article for the *Journal* in 1999. After this abortive attempt, he gave up trying to reform scientific societies, let alone the unwieldy apparatus of the Academy.

Nonetheless, he defended them in his own way, by resisting the proliferation of small scientific societies. The late 1980s, for example, saw the creation of the *Académie européenne, the International Academy of Science* and *Academia Europaea*, all of which sought to entice him onto their scientific committees, but he answered all three in the same way. "*I am sure that this spontaneous explosion expresses the best of intentions. I am not convinced that it will be very effective. My preference would be to merge a number of scientific societies and to construct an interdisciplinary body between them. (...) For the moment, I prefer not to participate*", he wrote to the founder of one of them in 1988. He also came out against the creation of a biophysics society in 1981. "*I believe in the importance of scientific societies. But I am also convinced that fragmentation is seriously damaging: 1) I believe that it is a bad thing for a young scientist to be confined within a narrow specialist field, where there will be no contact with new developments in other disciplines (...). This is an obstacle to learning and also mobility: all these specialist groups encourage lifelong confinement within one field. 2) If a young researcher wanted to react against the aforementioned deficiencies, he would have to maintain membership with one or even two other societies: three subscriptions is too much*", he commented. "*Excessive specialisation is damaging to scientific progress. It hinders interdisciplinary activity*", he asserted.

Boycott of *Nature*

On the same grounds, he also opposed the burgeoning development of scientific journals though he was constantly invited to join their scientific or editorial committees. As far back as 1977, he wrote on the subject of a planned journal on phase transitions: "*I personally have strong reservations about this project. (...) The growing number of journals on limited subjects mirrors one of the undesirable trends of our age, by encouraging excessive specialisation amongst the young."* In 1981, he also opposed the creation of a French journal on polymers, arguing that seven journals were already publishing articles on the subject. In 1989, he again responded negatively to plans for a journal of molecular engineering: *"[Your] proposed review tackles an important subject, but — on reflection — I am not in favour of it. We are seeing a disastrous growth in the number of journals (...). The effect of this growth is to inhibit the flow of information: a) it is physically impossible to read all the journals in one field; b) libraries are running out of funds (...); c) with this system, a great number of mediocre articles get published (...). For your case, all the necessary channels already exist. (...) The only person who clearly profits from such a project is the publisher. It is a combination of the lure of money and the naivety of certain scientists that causes the problem. In any case, if a new review is needed in a particular sector, my clear preference is that it should be run by a scientific society — and not for profit"*, he explained. On the other hand, he did support the launch of the European journal *Soft Matter,* eventually published in 2005 by the UK's *Royal Society of Chemistry.* He wrote an article for its first issue entitled *"Soft matter: more than words"*, in his capacity as the founder of the science of soft matter. *Soft Matter* is now a leading journal in its field.

Unlike the majority of researchers, Pierre-Gilles did not always seek to publish in the most prestigious journals, despite the fact that they would always have made space for him. For example, he refused to publish in *Nature* after being asked to write an article on polymers that never got published, because the subject had eventually been dropped. *"I have no desire to relive that experience"*, he replied curtly, when declining the request from *Nature* to write for the journal again in 1983.[5] By contrast, he became a stalwart defender of the *Proceedings of the Academy of Sciences* (despite his feelings about the institution). He published more than 170 articles there (many in French). *"He is the undisputed world champion for published articles in the Proceedings!"*, in the words of Paul Germain, Permanent Secretary of the Academy of Sciences (from 1975 to 1995). And that was not all,

---

[5] Not until 2001, when he would publish an article there on ultradivided matter.

since he also convinced eminent American physicists, such as Michael Fisher, to publish with him in the *CRAS*. *"I was proud to see this great representative of statistical physics in a French journal"*, he averred.

By publishing in the *Proceedings of the Academy of Sciences,* he ran the risk that his articles would go unnoticed, since the journal's circulation outside France was small. This is evidenced by a misadventure in 1998: reading a review article on wetting, he noticed that neither he nor his collaborators had been cited and was furious. *"I must confess that I had a shock this week on (fortuitously) reading your review on wetting, where most of the work of French research teams on the dynamics of wetting is not cited. (...) You must have very good reasons not to mention our work on layered droplets with A. M. Cazabat (...)"*, he wrote to the German physicist who had authored the article. The latter responded: *"The* CRAS *are scarcely available outside France and I must confess that I was not aware of this work. (...) I firmly believe that if a scientist has made findings that he considers truly important, he should publish them in English, to make them accessible to as many readers as possible."* Pierre-Gilles was well aware of this, but nevertheless did not change his methods.

Scenes at the CNRS

In 1980, a year after his election to the Academy of Sciences, he won the CNRS gold medal, France's most prestigious scientific distinction. He did not boycott the medal ceremony. *"I went there with my mother, very dignified with her walking stick* (Author's note: she was now 90). *The minister, Alice Saunier-Seïté, was very kind to her. She was an energetic and picturesque woman, very sporty — she used to fence on the roof of her ministry — I liked her a great deal"*, he recalled. All the leading lights of French research were there. In his acceptance speech, the physicist took the opportunity to make a point. He spent no time on thanks and the other typical niceties, but, having pointed out how aware he was of being *"interchangeable"*, he criticised the threat to research from the *"growing civil-service ethos"* and explained: *"[The researcher's] advancement, decided by elected and unionised committees, will be primarily based on length of service. I know several pioneering scientists who have been kept at the bottom of the ladder in this way. Let's not beat about the bush — if this approach continues, research is doomed".* According to him, *"we must protect job security, but not jobs for life".* He then proposed that CNRS researchers *"who are running out of steam"* should spend a few years teaching into the *classes préparatoires,* whilst teachers should come into contact with research and academics with industry. He was promulgating *"cross-fertilisation"* between the different scientific disciplines. Revolutionary!

He had never hidden his beefs about the CNRS, which went back a long way, to the time when he sat on his first CNRS committee, in 1963, at the age of 31. The chairman of the committee was all-powerful, which infuriated Pierre-Gilles. Later on, he would claim that his scientific authority gave him no special influence over decisions, and the commission often even voted against his recommendations. Used to his word being law, he was very put out and became convinced that decisions were not always taken on scientific grounds. In 1969, for example, he was part of a committee charged with deciding the future of a research lab at Meudon, near Paris: *"This lab had been losing ground for years and needed shutting down. But there was too much cronyism and union influence",* he explained. In fact, the commission voted for a motion of encouragement... and Pierre-Gilles slammed the door: *"After May 1968, power shifted from the mandarins to the unions",* he complained.

After this, he refused to spend time assessing labs and researchers, and never again set foot in a CNRS committee, instead complaining frequently about their decisions. *"On reflection, I feel it my duty to write to you about the appointment of research directors in your section. It seems to me that the rejection of [this researcher] is scientifically scandalous. He is a world leader in polymer science and interface science. (...) The committee's decision has driven him into the arms of industry. It has destroyed all the efforts that [I and others] are making to keep him at the CNRS. In the face of such a disastrous decision, I no longer feel inclined to defend the CNRS ",* he wrote to a committee chairman in 2002, with a copy to CNRS management. *"When I see admission juries making such arrogant decisions, I sometimes wonder if the jury is as good as the applicant...*", he complained privately.[6] He also had problems with the compartmental nature of the committees. *"The [Orsay submolecular biophysics] team currently depends on Committee 19 (macromolecular physical chemistry), but is somewhat marginal to the interests of that committee. In particular, the theoretical post-doc researcher (...) is currently without a job (...). This is a shocking situation, which illustrates how hard it is for biophysics to survive, with the lack of recognition within the CNRS"*, he wrote in 1978 to the director of biology at the CNRS. Nor was he satisfied with the choice of subjects, which *"does not always clearly reflect a long-term strategy. (...) The scientific poverty of the proposals shocks me. Personally, I am quite discouraged: several in-depth topics (colloids, non-Newtonians) to which I had committed myself and which require sustained action (...) currently seem to me to get little support at the CNRS"*, he complained in another 1978 letter, this time to the director of physics at the CNRS.

As with the Academy of Sciences, he was accused of carping without trying to do anything to reform the CNRS. *"I could probably have done more"*, he conceded, *"but it would have taken a lot of time"* — time he preferred to spend on research. In any case, he wasn't sure he could change anything. However virulent they were, he only expressed his critiques internally, and was careful to defend the CNRS to journalists and industrialists. When Philippe Jaffré, then head of Elf-Atochem, bluntly asked him during a lunch: *"Monsieur de Gennes, explain whether we really need a CNRS in France"*, he found the right words to support it. Moreover, while he denounced the failings of the French system, he didn't see the American system as a model. *"Young, untenured professors are under pressure: they need to get ahead at any cost, they take on too many students, they publish too quickly and they fight among themselves. I've seen a senior US scientist criticise their system and reach the conclusion: "What we need is a small, permanent body of young researchers, who could conduct bold experiments without fear of failure." He was reinventing the CNRS."*[7] Here again, his ideas are still

---

[6] Correspondence, May 4, 1998.

[7] *Commentaire*, 108.

relevant at a time when there are plans to break up the CNRS into different institutes.

Hit-Parade and Legion of Honour

In the early 1980s again, he became the French physicist most cited in scientific articles, as he learned from his former teacher, Anatole Abragam. *"At an age when the satisfactions of vanity increasingly take precedence over all others, intellectual or physical, you can imagine how pleased I was by the review of our book, which you were kind enough to send me.*[8] *Even more than the favourable assessment, what pleases me the most, though it comes as no surprise, is the lucidity of the analysis. (...) On a completely different subject, you may be amused to learn, if you didn't already know, that on the hit-parade of the Citation Index, for what it is worth, where for 15 years my books ensured me a place as the leading Frenchman, I was for the first time clearly overtaken in 1981 by a certain Pierre-Gilles de Gennes (who had long surpassed me in other, more interesting respects)"*, reads a letter from Anatole Abragam in 1982.

After his election to the Academy of Sciences and the CNRS gold medal, the next venture onto the red carpet was the *Légion d'honneur* in 1989. It was his eldest daughter, Dominique, who answered the phone call at home in Orsay: *"Daddy? Chevalier de la Légion d'honneur? But he'll never accept it."* Her reaction made him laugh. In the end, he received the decoration from Lionel Jospin, Minister of Education, in 1990. Subsequently, he was also named *Officier* in the 1992 New Year's list (at the same time as Philippe Nozières). This time, he allowed 11 years to pass before receiving the distinction, postponing the chore of a ceremony, but earning himself a dressing-down in the process. *"As you know, [two researchers] have expressed the wish to receive the insignia of Chevalier de la Légion d'honneur at your hands. I have the honour to inform you that I am not currently in a position to issue you the documents required to make the award to the interested parties. Indeed, following your promotion to the grade of Officier de la Légion d'honneur (...),*

---

[8] The book was *Nuclear Magnetism, Order and Disorder*, jointly written with M. Goldman and published by Oxford University Press in 1982.

*my office sent you a letter specifying the formalities to be accomplished (...).
Since no reply had been received to this letter, a further communication was
sent to you on February 22, 1994, which likewise went unanswered",*
he read in September 1998. In 2003, he decided to ask his cousin,
the neurophysiologist Yves Laporte, who was a director at the
Collège de France, to make him an Officier *"as simply as possible",*
and the latter organised the handover of the decoration in his
office, with no other witnesses.

More Honours

Pierre-Gilles de Gennes also received recognition outside France.
He was appointed Professor-at-large at New York State's Cornell
University in 1977, hearing the news from his friend Vinay
Ambegaokar. *"Around four years ago, I told you about the Professor-at-large
program at Cornell. Since then, your friends here (Wilson, Fisher, Nelkin,
Mermin, myself, etc.) have been working to get you appointed. I have just
heard that we have succeeded",* he wrote.

The two men had known each other since 1963, when the
Frenchman wrote to the American to draw his attention to a calculation error. *"I would like to emphasise that my remark concerns a minor
numerical error and that, apart from this, I entirely agree with the general tenor
of your article".* The researcher's straightforward reply — *"You are
absolutely right!"* — appealed to Pierre-Gilles, who subsequently told
him that he had spotted the mistake because his intuition told him
how the current should depend on the temperature in a certain
regime, which was not reflected in the final equation: he had then
gone through the demonstration with a fine-tooth comb, until he
found the calculation error. The two men would later do research
together, notably on superfluid helium, a very low temperature phase
in which helium flows with zero viscosity (A10-1).

*"I really enjoyed Cornell. It's very beautiful in spring — not in winter
when there are tons of snow, or in summer when it's hot and humid. It's a
great place for science."* Indeed, Cornell University proved to be as
good a place to meet people as he had hoped: he rubbed shoulders
with the top physicists on the campus, men like Ben Widom,

Kenneth Wilson, David Mermin, etc., and worked with some of them. One of these was Michael Fisher, a phase transition theorist. They worked on a problem of phase transition in a fluid near a wall, for which the French physicist proposed a simple scaling law. *"But Fisher didn't believe it, because he had looked at complex cases which produced new exponents. So we went over it again on the board and he accepted that these bizarre new exponents did not appear in our case, and that I was right about my scaling law. It was on this occasion that we jointly published the short article in the* CRAS *that I'm so proud of "*, he explained.

At Cornell, even the French theorist's lunches were productive! One morning, he spotted Roald Hoffmann (Nobel prize for chemistry in 1981) in the queue at the Faculty Club where the teachers ate, and started up a conversation: *"Why is organic chemistry carbon-based? Would it be possible to have a similar silicon-based chemistry? "*, he asked. Roald Hoffmann took his time before answering: *"When the orbitals are full, two silicon atoms cannot get as close together as two carbon atoms, because their orbitals don't overlap so well. "* Pierre-Gilles was silenced by the beautiful simplicity of this answer. He would return from Cornell galvanised by these encounters. *"When you compare the vitality of Cornell or Berkeley with Paris, it's worrying"*, he would say every time he got back.

Closer to home, he received honours from the British scientific community with his appointment, in November 1984, as a member of the Royal Society (the equivalent of France's Academy of Sciences). He had to sign the members' book, whose eminent signatories included Isaac Newton, at a sumptuous ceremony, which he attended this time impeccably attired. A few days before, he had experienced a misadventure at the dinner organised in his honour by the illustrious British physicist Sam Edwards, at Cambridge University's prestigious Gonville and Caius College, and it was still fresh in his memory. The taxi taking him there had gone off the road and ended up in a field. Pierre-Gilles had to wade through mud to get out of the car. A farmer with a tractor had towed the vehicle out, and the French physicist has finally arrived at the dinner on time, relieved not to be late, but... embarrassingly dishevelled.

Other distinctions included the Matteucci Medal, awarded in 1987 by the Italian Society of Sciences, for work in fundamental physics, the Harvey prize, awarded in 1988 by Israel's Technion Institute, and then the prestigious Wolf prize in 1990. *"It has become more than a rumour that you have won the Wolf prize. It is a choice that honours the prize, rather than vice versa (...)"*, wrote Phil Anderson, whom Pierre-Gilles saw at the time as a leading figure in solid-state physics. *"Thank you for your message. Encouragements from a guide like you are highly stimulating"*, the winner wrote back in thanks. *"His measurements are 6' 4"; head circumference: 22 inches; sleeve length: 24 inches"*, indicated his secretary in preparation for the ceremony in Jerusalem. He then received the Lorentz medal from the Dutch Royal Academy of Arts and Sciences in 1991. *"I feel very intimidated to feature in this very impressive list of winners"*, he replied. All these prestigious prizes would pave the way to the future Nobel Prize.

In 1991, he was also awarded an honorary doctorate at Cambridge by Sam Edwards. The pageantry of the awards ceremony was impressive. The winners, in black gowns and red mortarboards, were assembled, then ushered in rows before the Duke of Edinburgh, the Queen of England's husband Prince Philip, who was seated on a throne wearing a blue train several metres long. One after another, the speakers stood up to introduce the postulants in Latin. Pierre-Gilles became bored. Noticing an attractive woman beside him, he began a conversation. *"Are you from London?"*, was his opening gambit. She looked at him in surprise. *"No, I'm not from London. I am the President of the Republic of Ireland"*, she smiled gently. Mary Robinson was also being honoured that year. After the ceremony, still in gowns, they paraded through the streets of Cambridge. Pierre-Gilles was apprehensive, fearing an IRA outrage against the Duke of Edinburgh. *"The situation in Ireland at the time was tense. There were many Irish families cheering and waving flags on the route, but I was thinking that there could also be a guy with a bazooka on a roof"*, he admitted.

One Foot in Industry

Industry also joined the bandwagon, offering the French physicist huge incentives in return for his services. *Exxon* engaged him as a consultant in its research centre in Annandale, New Jersey. Set up in

1980, it had recruited eminent physicists like Phil Pincus, John Schrieffer (one of the authors of the BCS theory), Tom Witten, the polymer specialist who would spend a year at Collège de France, the rheology specialist Bill Grassley, and many others. Pierre-Gilles spent a week every year there, impressed by the ambitious research projects underway at the centre, but the partnership ended badly in the late 1980s. Arriving at the centre, he was met by a manager who challenged him sharply: *"What are you doing here? You didn't come at the scheduled time, so you have been dismissed. The management wants you to reimburse what you have been paid."* Pierre-Gilles was taken aback, but nevertheless completed the schedule he had planned for the week, left without making a fuss, but of course never returned. *Exxon's* management would kick themselves when he was awarded the Nobel Prize three years later, and vainly tried to entice him back into their fold.

He was also in demand in France. In 1987, he was approached to become scientific director at *Rhône-Poulenc* (split into *Rhodia* and *Aventis*, now *Sanofi Aventis*). He declined the offer, explaining that he could not be available more than one day a month, and suggesting other possible candidates. However, *Rhône-Poulenc* wanted him. So he accepted a position as a consultant, then as a member of a trio of research directors with Jean-Marie Lehn (Nobel Prize for chemistry in 1987) and Claude Hélène (a biologist and member of the Academy of Sciences). In reality, he worked in tandem with one of the company's executives, Jean-Claude Daniel. Once a month, the two would meet at 7 a.m. at the Gare de Lyon, and take the train to Lyon. During the journey, the theorist would learn about problems emerging from the labs, then visit workshops and take part in conferences. He would meet around forty teams a year and helped to introduce a number of measures. For example, he and Françoise organised a continuous training course for the company engineers. He also worked to unite the academic and industrial communities, by backing the creation of two joint labs,[9] by encouraging the recruitment of physics PhDs into industry and devoting a session of the Les Houches Summer School to industrial issues in 1998. *"With Rhône-Poulenc, I feel I was of some use, in fact,*

I think I contributed to a change of attitude within the company", he concluded.

As the offers came thick and fast, he had to make choices. In 1989, he became a consultant at a start-up called *Flamel*, out of friendship for its founder Gérard Soula, a man seething with ideas (antibacterial gloves, delayed insulin, etc.). He tracked this company's progress enthusiastically (at one point it employed 50 researchers) until it was taken over and his friend stepped down. But he declined other offers, for example to become a member of an ESA (European Space Agency) think tank in 1985, or consultant to the UK firm *Unilever*. Sometimes, his opinion was not even asked... For example, he discovered in 1990 that — on the recommendation of several important figures — he had been summarily appointed to sit on the Senior Scientific Committee of the Paris-Meudon Observatory. They had just forgotten to ask for his agreement! *"The procedure just went ahead"*, they explained. He was not amused.

Lab Groupies

Between the different consultancy roles, and the classes and lectures he had to give, plus the directorship of one of France's higher education institutions from 1976 (Paris Higher School of Physics and Industrial Chemistry),[10] he was more often on the move than in his lab at the Collège. He relied entirely on Madeleine Veyssié, whom he appointed director, and Christiane Taupin (followed by others) to manage day-to-day problems. They dealt with all the administrative hassles. *"He never even saw an order form or a business report. We arranged things so that he devoted the little time he spent at the lab to scientific tasks"*, recalls Madeleine Veyssié.

One factor that made the lab stand out was its large female contingent, a fact rare enough to attract comment. As well as the lab

---

[9] A joint *Rhône-Poulenc*, CNRS and Princeton University lab, working on tensioactives and water-soluble polymers (complex fluids), and joint CEA and *Rhône-Poulenc* lab in Aubervilliers working on suspensions.
[10] See Chapter 12.

itself being headed by a woman, most of the teams also had female leaders, for example Liliane Léger, Claudine Williams or Anne-Marie Cazabat, mockingly nicknamed by some the de Gennes *"groupies"*. Was this strong female presence an accident? It's unlikely. Was it because Pierre-Gilles attracted them — *"They were all in love with him"*, claimed the mockers again — or because men kept away — since the brilliance of a Pierre-Gilles de Gennes could undermine certain inflated male egos in the research community? According to the women concerned, there was simply genuine gender equality in the lab, with the result that female physicists were more visible and more comfortable there than elsewhere.

Although he was often away, Pierre-Gilles was not an absentee boss. When he was in, the door of his office was always open. *"A lingering smell of cigarillos in the corridor signalled his presence"*, recalls Madeleine Veyssié. And like a wisp of smoke, the news would spread around the lab. Booboo, who was strategically located near his office, was in charge of notifying colleagues on the other floors. *"He's here"*, she would simply say on the phone, and people would understand. Anyone who wanted to see him was accommodated immediately, or at least the next day. Keen to put people at ease — *"especially youngsters"*, stresses Liliane Léger — Pierre-Gilles was easy to talk to. He would listen intently and look you straight in the eye. However, younger researchers, postgraduates or postdocs, did not always have the courage to knock on his door, although they had everything to gain in talking to him. Those who overcame their nervousness would inevitably emerge from the discussion stimulated by the immense enthusiasm the physicist communicated. *"One remark from him could save six months of research"*, comments one of his students. *"The things I learned from you are going to help me for a long time to come. What I liked most of all was our conversations at the blackboard, where I got a sense of how your mind works"*, a young US researcher wrote to him after a stay at the lab in 1982.

In addition to the open door policy, Pierre-Gilles kept closely informed of events in the lab and the progress of everyone's work. He would often invite one of his colleagues to a restaurant in lunch breaks. *"We need to talk. Are you free midday Wednesday?"*, he would ask, sticking his head round the office door. *"A crucial part of the life of the lab went on, and was even decided, at these lunches"*, acknowledges Madeleine Veyssié. He would canvass opinions, listen to complaints, track progress. Nothing of significance escaped him.

And finally, when he was around, he would never miss a meeting or a lecture. *"When he was there, people might be a little more careful about the questions they asked, for fear of missing the obvious, but his presence would never strangle a meeting"*, stresses Marie-Alice Guedeau-Boudeville, "Booboo". The discussions would always take place in his office, which as the lab's biggest room was used as the seminar room. The senior staff would sit in the armchairs, the others on chairs and latecomers on the floor. He himself would sit at his desk, opening mail. The noise of envelopes being opened could be heard above the speaker's voice. He would say little. It looked as though his mind was elsewhere, but this was a false impression, because when he did intervene, his comments went straight to the point. He would follow the demonstration while going through his correspondence. *"He was capable of assessing whether an equation was right within a few seconds, simply by analysing its form"*, one researcher insists. He was also able to follow the argument while devising a model that included what was being said and the next day would send his calculations to the speaker, with a little note simply saying *"What do you think?"* Finally, his presence at the laboratory, however occasional, was intense. Even when absent, he left his mark — his attitude and his enthusiasm — so that everyone worked hard to achieve good results and prove their worth.

Credit Control

He was not just an eyes-up lab director, he was hands-on. Although he delegated many tasks, he made the strategic decisions almost single-handed. He would listen to the views of the senior researchers, a sort of unofficial think tank, and then make his decision. For example, at the crucial funding allocation meetings, which took place in spring, team leaders had 15 minutes — not a minute more — to present their accounts and budget requirements. During the process, Marie-France Jestin, the secretary, would fill in each team's column on the board. There would be lively discussions, as the different researchers made their pitch. Pierre-Gilles would listen and then clear his throat, the sign that he was about to bring negotiations to an end. Silence would fall. He would allocate the amounts to each team. No one would say a word. The meeting would be over.

He also made all the decisions on research. One day, he announced his intention to begin research on wetting. There were glum looks at the idea of abandoning ongoing experiments. Irritated, he snapped: *"If you want to do academic science, fair enough, but it won't be with me."* The die was cast, and the wetting research he wanted began soon afterwards, as researchers agreed to end their current projects, making a sacrifice of which Pierre-Gilles was well aware. *"When an experimentalist changes subject, he leaves his community. His expertise is no longer of any use. He has to start from scratch, step up to the mark again"*, he would later acknowledge. Here, he proved himself as demanding on his collaborators as he was on himself. As a sworn enemy of stagnation, he could not tolerate it within his own laboratory.

In fact, he also encouraged mobility amongst researchers. As they acquired experience, they might want to develop their own ideas and build up a team. But the lab was small and Pierre-Gilles refused to request more space. In fact, he took advantage of the lack of room to prevent teams proliferating, like a corset keeping the laboratory lean. *"You don't measure the quality of a lab by the number of researchers and the surface area"*, he would say to anybody who complained. So sometimes as many as five people would be crammed into a single office, and starting a new experiment meant dismantling an old one. Those who wanted to spread their wings would move elsewhere. Once students had completed their PhD, they had to leave the fold, except for a few who were recruited, such as David Quéré and Élie Raphaël. Ultimately, it was less a laboratory for people to develop their own subjects than a place designed to "exploit" the boss's grey matter, though this does not mean that the researchers did nothing but check his predictions — far from it, since their experiments were also the inspiration for many of his models. The collaboration worked synergistically.

Yvonne de Gennes in
front of Château de
Sceaux, August 1933

Pierre-Gilles aged one,
in the family apartment,
Avenue de Camoëns,
in Paris

Antibes, 1936

Avenue de Camoëns

With a nanny, 1935

In the family apartment, 1935

October 1937

Excursion to Gennes, May 1938

November 1938

With his mother, July 1938

Trouville, Spring 1938

With his father, October 1939

With his cousin, on his sixth birthday,
24 October 1938

Barcelonnette, December 1942

With his mother, at Villard-de-Lans,
December 1940

Class second B at lycée Claude Bernard in Paris, 1947 (left to right, Pierre-Gilles is the 2nd student on the back row. Bernard Prugnat is the 2nd in the front row and Bruno Schroeder the 3rd. Jacques Chemin is to the right of their maths teacher, Camille Lebossé).

UNIVERSITÉ DE PARIS

École Normale Supérieure

45, Rue d'Ulm, PARIS

Carte d'Élève de l'Ecole, valable pour _1_ an
à dater de l'année de la promotion 1953
Prolongée pour l'Année Scolaire 19

M _de Gennes Pierre Gilles_
né le _24-10-_ 1932
à _Paris_

Le Titulaire de la Carte,

P. G. de Gennes

Le Directeur de l'École,

With Pierre Favard, 1954

In the Queyras, 1952

Marriage at
the Temple
de Neuilly,
June 1954

Conference on magnetism in Strasbourg, July 1957 (Pierre-Gilles is 4th from the
left starting from the bottom; from left to right, Jacques Friedel is 3rd in the
back row and Bernard Jacrot is just to the right of him; in the front row,
Cornelis Jacobus Gorter is 3rd, Louis Néel is 5th, Edmond Bauer 6th and Louis Weil 9th)

Pierre-Gilles with Maria, his father's governess, and Annie,
in Gloversville in the USA, September 1959

In Orcières with Annie, 1973

1971

1975

1979

1979

1991

1995

1972

Sketching,
1978

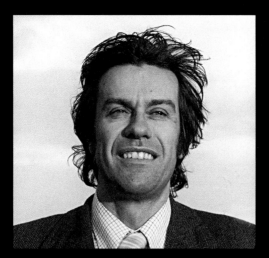

1972

1991

Receiving the
American Chemical
Society prize, with
Françoise, 1988

Pierre-Gilles surrounded
by his friends at the
Nobel Prize award in
Stockholm, 10 December
1991 (from left to right,
Schlomo Alexander,
Étienne Guyon, the
laureate, Phil Pincus,
Madeleine Veyssié, Karol
Mysels and Jean Jacques)

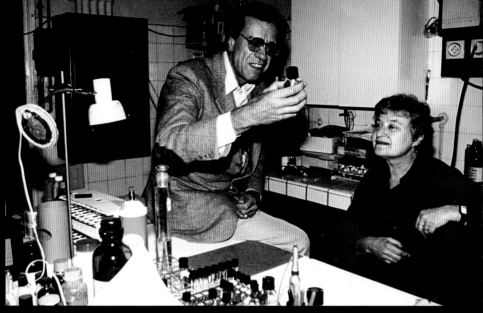

With his friend, the chemist Maya Dvolaitzky,
at the Collège de France laboratory, 1991

Lecture on liquid crystals in Rio de Janeiro, January 1972

In his office at the Collège de France in the 1990s

At ESPCI in 2002

So Pierre-Gilles kept his lab at a size that he thought conducive to the flow of ideas and to friendly relations. In fact, there was a family atmosphere, which was good to work in. *"I have never been welcomed as a visitor with so much warmth and hospitality as in your lab"*, wrote one physicist after a stay. The coffee jug, located in a corner opposite Pierre-Gilles' office, was the focal point of the lab. Sometimes, leaning against the wall, the theorist would chat over a coffee, especially with the chemist Maya Dvolaitzky, who had become a friend. The doors of all the offices were open. In the spring, there would be a picnic, and at Christmas, lunch in the laboratory corridor. Everyone would bring a home-made pie or salad, a door on trestles covered with a paper tablecloth would serve as a table. Maya would provide burgundy and Pierre-Gilles champagne. Sometimes, after the meal, the party would continue with a little dancing.

Trickle of Students

Pierre-Gilles was very keen that his lab should not become a school for theorists, so he would only supervise one doctoral student at a time. After Saad Daoudi, in 1977, he took on Jean-François Joanny, a reserved young man who had approached him against the advice of his teachers at the École Normale Supérieure. According to them, Pierre-Gilles de Gennes was a hands-off supervisor, who didn't spend much time on his students. But Jean-François Joanny had stuck to his guns. He had been absolutely determined to work with him, since attending one of the physicist's lectures, which was *"unorthodox both in style and in the subject he covered."*

Pierre-Gilles welcomed him warmly, suggesting other potentially suitable labs, then sent him to see Mohammed Daoud, a theorist at CEA Saclay, with whom he was working on polymer research, saying: *"He can tell you better than I can what it's like working with me."* *"Don't expect to have your hand held, but go for it! You'll never meet another physicist like him!"*, was the advice. Another student was Élie Raphaël, whom Pierre-Gilles supervised from September 1987. A graduate of Strasbourg University, he had been warned: *"You haven't been to a grande école, so you'll find it difficult to do a PhD in theoretical physics."*

Instead, he had the chutzpah to approach Pierre-Gilles de Gennes, who welcomed him with open arms with no other formality than a friendly interview. After that, there was Cyprien Gay, Achod Aradian then Thomas Boutreux, but all in all he had few PhD students.

For all of them, his style of supervision was similar. First, as a vacation assignment, he would prescribe several books on subjects as varied as superconductors, polymers or liquid crystals, and then make them work on... something else. Postgraduate students would discover their actual research subject at the beginning of the academic year, over lunch, where Pierre-Gilles would cobble together a class on their future research. *"I remember a meal with him in a Chinese restaurant, right at the beginning of my research. He was explaining his idea, as usual scribbling equations and diagrams on the paper table cover with his critérium pencil — he always used a basic plastic critérium (he liked looking at beautiful things, but couldn't care less about having a sophisticated pen or watch). I was beginning to sweat blood, wondering how I was going to retrieve that piece of paper. In fact, he might have found it funny if I had picked it up when leaving the restaurant..."*, recounts Élie Raphaël. Then, the physicist would leave his students to do their own thing, and sometimes mess up, *"but only just enough for them not to lose their way"*, he smiled. He would regularly check their progress, offering explanations and encouragement. *"It takes 15 years to become a researcher"*, he would repeat. He was no longer as cutting as he had sometimes been in his early days as a supervisor.

Many of his students inherited his style. Like Pierre-Gilles, they collaborate with experimentalists. Like him, they look for relevant approximations, etc. *"You taught me to see simplicity in complexity"*, one young researcher wrote. Jean-François Joanny is even seen as his "spiritual heir". *"He has the same curiosity, the same reliable physical intuition, an exceptionally economical way of elucidating a new phenomenon"*, one eminent French physicist said of him. *"He inherited from de Gennes a style, a very characteristic "hand", which takes him straight to his goal."*

However, having Pierre-Gilles as a supervisor was not an unalloyed advantage. *"Scientifically, he is so clever and quick that he leaves any young theorist behind, however brilliant"*, opines one physicist. For example, one day when a colleague asked him to help resolve a theoretical problem regarding a crystalline structure, he replied: *"I don't have time to do that this weekend. But it would be a good subject for a doctoral student."* Sometimes he would give the experimental team the results of calculations that he had asked his student to do, advising them not to say anything about it, because *"he needs to do the calculations himself"*. At other times, he would be less discreet. Once, after asking Jean-François Joanny to perform a calculation (on the statistical behaviour of a polymer mix), he finally did it himself, since he needed the result for his class at the Collège. Attending the class, the young researcher was stunned to see the very demonstration that he had been sweating over for days! It was only afterwards that Pierre-Gilles realised his blunder. His talents could sometimes make researchers feel like a spare wheel, which could be discouraging, even crippling. A sort of giant tree casting a shadow over all the young shoots.

There was another downside to working with someone of his stature. He was like an oasis in the desert — an inexhaustible source of ideas — and it could be hard for people to quit to follow their own path. *"It is true that there's a risk of becoming dependent, of becoming a mere instrument"*, agrees Christiane Caroli, his student in the superconductor period. And even when his collaborators were anything but instruments, they were still accused of being so — another far from insignificant pitfall of working with Pierre-Gilles. Indeed, they were inevitably suspected of having done no original work themselves. However much the theorist denied it, doubts remained. Françoise was the first to suffer, but also Élie Raphaël and Jean-François Joanny. When Joanny and Ludwik Leibler, then working as a post-doc in the lab, discovered the phenomenon of polymer depletion at a wall (the polymers are repelled from the wall when they have an affinity for the solvent), Pierre-Gilles fought hard to ensure that his students got full credit for this discovery. However, he generated so many ideas so quickly, that it is impossible to know what came from him and what did not.

Unfulfilled Hopes

Ultimately, Pierre-Gilles de Gennes succeeded in setting up a productive experimental laboratory at the Collège, capable of supporting him in his new research topics. However, he failed to bring in the multidisciplinary research he had wanted. True, he managed to make physicists and chemists collaborate on biological cells, focusing in particular on how they merge, getting Maya Dvolaitzky and Booboo to synthesise "giant vesicles" made up of a phospholipid bilayer, simplified models of biological cells. The chemists then confirmed the laws that they had predicted on the pressure conditions that promote cell merger (A10-2). However, there was no partnership between this research and the Collège's biology labs. Pierre-Gilles was disappointed that the disciplines remained compartmentalised, unlike their English counterparts. Even the synergy with the chemists was not as effective as at Orsay, since Maya and Booboo had to focus on just one of the many fields covered in the laboratory.

Though he failed to initiate research at the cusp of physics, chemistry and biology, as he hoped, Pierre-Gilles instead invented soft matter physics, establishing his reputation in France and even internationally. *"I have long thought that your research on liquid crystals, polymers and flows — closely tied in with experiment — shows how condensed matter physics can remain a young and vigourous science"*, an eminent Harvard physicist wrote to him in 1983. For a long time, scientists had seen these subjects as trivial and uninteresting technical problems. *"How can such an intelligent lad get interested in soap?"*, Anatole Abragam had jokingly asked him. And indeed, he had to struggle to promote this new physics, for it to gain acceptance both in attitudes and in the teaching curriculum. Contacted in 1986 to give his opinion on the content of the physics Magistère degree at Orsay, he did nothing to hide his disappointment. *"Personally, given the lack of interest shown by Orsay in subjects that I perceive as important (colloids, polymers, fluid interfaces, random media), I do not think that it would be of any use for me to get involved at this late stage"*, he replied. As late as 1997, consulted on a short introductory article on physical chemistry, he was deeply disappointed. *"The work on polymers, liquid crystals, colloids, interfaces, wetting and adhesion, lipid-water systems, etc. — which is, it seems to me, a significant aspect of this field, receives absolutely no mention! (...) I'm afraid that what I see in your text is the reflection of a slightly old-fashioned view — primarily concerned with spectroscopy and theoretical chemistry — that still persists in France"*, he commented. In this respect, he is slightly reminiscent of the chorus of the poem by Paul Fort, *Le petit cheval blanc*, sung by the French folksinger Georges Brassens: *"Everyone following, everyone following, him alone at the front."*

Chapter 11

The Invention of Soft Matter

One day, Babar was kidnapped by aliens and taken to a strange soft planet, where he sunk into the ground as he jumped out of the alien spaceship. When, in the 1990s, Françoise unearthed the book *Babar Visits Another Planet* for her son Marc, she was struck by the analogy with the new science of soft matter, especially as the book had been written in 1974, well before the emergence of this new field of physics. It was a coincidence that also amused Pierre-Gilles, the primary craftsman of soft matter physics, which — beginning with polymers — was the focus of his research at the Collège de France.

In the early 1970s, polymer physics was going nowhere. The only case that had been properly described was that of a few polymers isolated in a solvent (called a dilute solution), like a few strands of angel hair floating in a soup dish. Here, it was as if the monomers moved randomly, one in one direction, the other in another, and so on, like a drunk emerging from a bar, except that these elementary entities could not form layers and repelled each other (the so-called "excluded volume" interaction). As a result, a polymer formed of a chain of multiple monomers ultimately looks like a swollen coil. The characteristics of this coil, for example its size relative to the number of monomers, were known and followed a power law (in the form $y = a.x^k$ where k is called the critical exponent). Things became more complex when there were more polymers in the solvent (in a so-called "semi-dilute" solution): they overlapped, screening each other, with the result that the repulsions between monomers were screened. The classical model (based on mean field approximation, see A7-9) produced a power law that did not match the experimental evidence (the exponents were wrong). The same was true of highly concentrated polymer solutions ("molten" polymers). The theorist Paul Flory (Nobel prize for chemistry, 1974) had explained the differences using approximate arguments. What else could he do? A precise description would have meant determining the relative positions of all the monomers, which was inconceivable. Physicists, Pierre-Gilles included, had no idea which thread to tug to untangle the knot.

"Chemists (...) bring us mysteries, in particular the mystery of polymers, those materials made up of long chains of molecules which, in the form of rubber or plastics, are currently transforming our environment for better or for worse. In most applications, the chains are entangled. In fluid phases, they are, in addition, capable of sliding over each other. One of the crucial problems then is to describe their movements: how does a massive knot of snakes form and dissolve? At present, we can't answer this. The best we have been able to do recently is to elucidate a smaller problem, that of a single snake moving randomly within an inanimate and motionless knot. But there is a considerable chasm between this smaller problem and the global problem", he explained in his inaugural class at the Collège de France in 1971.

A few months earlier, he had come up with a different approach: instead of considering the heap of polymers as a whole, he imagined a single chain trapped within the heap, like single strand of spaghetti within a congealed mass of pasta after a week in the icebox. He then imagined this chain sliding, very slowly, through the rigid obstacles, like a snake in a bamboo thicket, and from this managed to deduce simple behaviour laws based on work by the British physicist Sam Edwards. So significant was this "reptation" model, that all the laws that now govern polymer mechanics can be deduced from it, for example what makes one plastic bounce like a ball and another flow like a fluid. The reptation model achieved worldwide recognition. *"I was invited to China in 1988. The trip was tiring and wasn't going very well. At the end, I had to visit a nylon factory in Szechuan. And there I had a nice surprise when the chief engineer responsible for stretching the fibres said to me: "So you're the reptation man." I was impressed: someone in a factory at the back end of China was familiar with the model!"*, he chuckled. Subsequently, he would go on to apply the model to polymers made up of mixed monomers, branched polymers, etc.

Less than a year after the invention of reptation, he made a further discovery on polymers, one that is seen quite simply as the apotheosis of his research. It all began in September 1971, when his physicist friend at Harvard, Paul Martin, sent him the preprint of an article on phase transitions by Kenneth Wilson and Michael Fisher of Cornell University. The article came at just the right time: Pierre-Gilles was fine-tuning his first lecture at the Collège de France, which he had decided to devote to... phase transitions. He immersed himself in the text and quickly understood why his friend was in such a hurry to get it to him: the article was completely revolutionary. The authors had removed an obstacle that for years had thwarted physicists in their attempts to describe certain phase transitions, by inventing a tool called the "renormalisation group" (Wilson would win the Nobel prize for this discovery in 1982) (A11-1).

After $E = mc^2$, $n = 0$

"I completely dismembered the article and immediately lectured on it at the Collège, at a time when even the word renormalisation was unknown to condensed matter physicists. So everyone who was anyone in statistical physics in Paris came to my first year class — it was nice — and to be honest, the course was useful. But above all, the article helped me to make progress on polymers..." In fact, the more he thought about it, the more he wondered whether the "renormalisation group" could not be used as a tool to solve the knotty problem of entangled polymers. He spent the December vacation in his chalet in Orcières, skiing in the morning and working in the afternoon. The day after Christmas, he shut

himself in his study to pursue his ideas. He began to apply the renormalisation group and, when he posited that a parameter n was equal to 0, a door suddenly opened. He continued, verified, and there was no doubt: he had succeeded (A11-2)!

His reasoning took took him... into another dimension. *"I realised that a parallel can be drawn between phase transitions in magnetic systems and in interacting polymers. I was aware of the fact that phase transitions become simple in four dimensions, and I also knew that the excluded volume effect becomes trivial beyond four dimensions. Perhaps there was a parallel? Since renormalisation techniques had worked for reducing the number of dimensions, I said to myself that there had to be a similar technique for polymers. And here, through an analytical extension, by positing n = 0, the problem was solved. It was a big thrill!"*, he recalled enthusiastically. So by applying the renormalisation group to polymer statistics, he established a correlation between phase transitions and the behaviour of polymers, from which he would rigourously deduce the laws governing polymers. He was aware of the significance of this discovery and keen not to waste a single minute. *"I returned to Paris hot foot, on December 27, to write a short article — I never finished my Christmas vacation."*

At the Collège de France, the lab was almost empty for the holidays. Pierre-Gilles could write his article undisturbed. *"I often start writing fairly soon after finding something, because writing helps me to think, to order my ideas: describing the starting point — what you know, what you don't know — is very useful. Generally speaking, I like writing, putting things down on paper, watching the steps emerge and seeing that everything fits together."* In a few hours, his article was ready. *"I sent it to* Physics Letters, *which I knew would publish quickly. It is a small, not very well-known journal, but that didn't prevent n = 0 making a worldwide splash"*, he commented.

In the end, Pierre-Gilles used the renormalisation group to build the foundations of a broad general theory of polymers, in any dilution. For semidilute solutions, he introduced a "correlation length", which represents the size of a small polymer coil called a "blob", likening the semidilute solution to a sort of fishnet made up of these blobs. He showed that, as the polymer concentration increases, the blobs become increasingly small (the mesh of the net tightens) (A11-3). At very high concentration, as in the case of a molten polymer, the blobs are the size of a single monomer. Here, paradoxically, the chains once again become ideal (there is no more excluded volume effect), and revert to a random walk, as the Nobel prizewinner Paul Flory had predicted (A11-4). In conclusion, Pierre-Gilles replaced the semi-empirical approaches previously applied to polymers, in particular by Paul Flory, with a rigourous approach that would generate simple laws, consistent with experimental evidence and independent of the nature of the polymer. If there is one equation that might be seen to exemplify the work of Pierre-Gilles de Gennes, as $E = mc^2$ is linked with Einstein, it would undoubtedly be $n = 0$.

Paul Flory did not greet this achievement as enthusiastically as other physicists around the world. His view was that the Frenchman had at most tweaked a few exponents that he had been first to come up with, nothing to make such a fuss about. Relations between the two men would always be cool at best. On one visit to Paris, Paul Flory had dinner with Pierre-Gilles, at the invitation of Lucien Monnerie. *"Yes, it was hardly a love affair"*, agrees the latter. *"It also has to be said that they had very different approaches. Flory worked at the molecular scale, whereas Pierre-Gilles de Gennes had a more global approach."* Their relations would become no warmer at subsequent meetings, and even turned icy at the *Faraday Society* colloquy in Cambridge, England in September 1979, where Charles Franck, an eminent researcher at Bristol University, was one of the speakers. *"Franck started saying that one of Flory's models, based on a switching phenomenon, was unacceptable, remarking: "It's lucky that telephonists don't have to work with switches like this, because it would be impossible to make a phone call." It was a scandal, almost like Khrushchev banging his shoe on the table at the UN in 1960: Flory stood up, shouting that it was intolerable that he should be treated in such a way, etc. That evening, there was a sumptuous dinner in a beautiful college hall. Just my luck, I found myself sitting between Flory and Franck!"*, recalled Pierre-Gilles de Gennes. He feared the worst, already imagining some new scene between the two physicists over dinner, with himself caught in the middle. Indeed, the meal began badly: Paul Flory didn't say a word to him when he sat down, as the young Frenchman was known to be a friend of Charles Franck's. In the end, Pierre-Gilles, mustering all the tact of which he was sometimes capable, managed to talk to the two men one at a time without referring to the quarrel, and Paul Flory relaxed. *"All the same, Franck was right"*, concluded Pierre-Gilles de Gennes.

"The art of the theorist"

The "n = 0" article marked a turning point, not only in polymer physics, but also in Pierre-Gilles' own career. Until then, he had essentially based his calculations on so-called molecular field (or mean field) approximations to estimate the interactions between the objects in the system he was studying (A7-9). In other words, he was a classical theorist in his way of approaching problems. Now, he had adopted a different approach, through so-called "scaling laws". With this method, he would firstly establish a small-scale law based on the parameters controlling the system (the main difficulty being to determine them correctly) and then reconstruct the scale above, and secondly solve a problem by a dimensional analysis (A11-5). The theorist compared this new approach with that of the Impressionists. *"Until Ingres, artists tried to reproduce details accurately, but partly as a result of the newly emerging art of photography, this attention to detail became a slightly old-fashioned technical problem, which could be solved by a photo. There were other things to explore, and that is what gave rise to movements like Impressionism"*, he explained to the Académie des Sciences. This approach made him twice as effective. *"He was perfectly capable of mastering complex formal processes, but he was ten times as quick in identifying the relevant parameters, the factors that would be relevant to the result, and in this way solved a plethora of problems that no one could work out"*, notes the physicist Michel Laguës. However, he had no wish to turn his approach into a doctrine. So when one of his doctoral students wanted to include in his calculations effects that he himself had ignored, he let him get on with it. A good thing too, because the student found a law that he had never thought of. *"Frankly, if I can avoid getting involved in tough calculations..."*, he joked.

The fact remains that, just as the early Impressionists were jeered, the scaling law approach attracted sharp criticism, with some researchers going so far as to describe Pierre-Gilles' work as... *"approximate"*. He understood these objections and recognised that his estimations were sometimes bold. *"Our whole approach (...), would need to be backed by detailed (...) calculations. Moreover, it suffers from extreme simplifications (...)"*, he admitted, for example, in a 1988 article. However, he remained convinced that the method is necessary. *"At the risk of generalising somewhat, it could be said that the art of the theorist in physics is to know just how far to go with simplification"*, he asserted at his inaugural class at the Collège de France. But some physicists would never come around to his point of view. *"He has the art of achieving very good results with bad methods. Because he guesses the result in advance, he makes sure he gets it"*, some continue to maintain.

The Polymer Golden Triangle

Pierre-Gilles de Gennes started experimental teams at the Collège de France, at CEA Saclay and at the Macromolecule Research Centre in Strasbourg to check his predictions, establishing a triangular collaboration that he called *StraSaCol* in a chronicle in the *Journal of Polymer Sciences*. *"The experiments to prove his models continued over 10 to 15 years, using polymers that were, so to speak, labelled so that their movements could be tracked. Most of them took place at Saclay and in Strasbourg, but the orchestrator was Pierre-Gilles de Gennes"*, explains Jacques Friedel. Other research started up around the world: the soft matter physicist's theoretical discoveries revived interest in polymer science.

In 1978, Pierre-Gilles used his time at Cornell University to write a textbook on polymers whose title, *Scaling Concepts in Polymer Physics*, stresses the importance of scaling laws. When Henri Benoît, Director of the Macromolecule Research Centre in Strasbourg heard the news, he was upset: *"He's writing a book, so it means he's going to drop the subject, as he did with superconductors and liquid crystals."* He was wrong: this time, the physicist would continue to work on polymers, though with a focus on unusual cases.

One day, at a conference in Maastricht, Andrew Keller, a polymer specialist at Bristol University, remarked on their extraordinary elastic properties. On his way back to the hotel that evening, Pierre-Gilles reflected on these properties: *"Polymers are so entangled, they inevitably form knots when stretched, knots that are probably very tight..."* This simple observation led him to research from which he developed a model of knots.

He also wondered how polymers thread their way through a narrow tube. When the tube is wide (with passive walls), they enter without the coils changing shape. As the diameter decreases, however, they have to stretch to penetrate in single file. He extended his ideas to different polymers and, in each case, calculated the characteristics of their progress through the tube. At a seminar in Varenna in Italy, he presented the case of star polymers to an audience of top theorists. He explained that the lowest energy configuration is found where half the polymer's arms face forwards and the other half backwards. At the end of the paper, Françoise asked a question from the floor: *"What if one arm faced forwards and the others backwards?"* After a moment's reflection, Pierre-Gilles could see no objection to such a configuration, which is even statistically more common. *"In a single remark, she completely demolished my paper"*, he remarked. Nonetheless, he was not in the least put out, as he welcomed any relevant comment.

In a few years, Françoise had become his chief scientific collaborator. At the Les Houches Summer School in 1953, Eugenia Peierls, wife of the physicist Rudolf Peierls, had said that scientists could be divided into two categories: tennis players and golfers — the golfer plays alone, striking the ball forward until he eventually reaches his goal, whereas the tennis player rallies with different players. Pierre-Gilles described himself as a "tennis player", and Françoise was undoubtedly his most frequent hitting partner.

Françoise, now also his partner in a more intimate sense, came from a family of scientists. Her railwayman grandfather had had two sons, one who studied physics, the other chemistry. Françoise's uncle was the eminent crystallographer Jean Wyart, a professor at the Sorbonne. Her father, Pierre, had built hydroelectric dams for

EDF, France's national electricity company, later becoming one of the architects of France's nuclear energy policy. On her mother's side, Françoise had a grandfather who was a chemist, and her great-great-uncle was none other than the chemist Charles Gerhardt, born in 1816, the first man to synthesise aspirin. Françoise had known Pierre-Gilles since taking his pre-doctoral classes in solid-state physics in 1967 as a student at the École Normale Supérieure in Cachan, after a faultless school career at the girls' lycée in Saint-Cloud and preparatory classes at Lycée Janson-de-Sailly in Paris. Almost by accident, she then found herself studying for a PhD with him. *"After my postgraduate research, I started a thesis on superconductivity at the LPS, at a time when everyone was switching to liquid crystals. After a year, I begin working with Pierre-Gilles on liquid crystals. I got stuck on the first calculation he gave me to do. So I looked at the orders of magnitude, then I went to see him, almost apologising for getting a result by estimating orders of magnitude. In fact, he found my approach interesting"*, she recounts.

Between 1970 and 2007, he would co-author 60 articles with her, twice as many as with any other collaborator (it should be noted that he was sole author of more than half his articles — the tennis player also liked to practise against a wall). They rarely worked together at the lab, caught up in too many day-to-day obligations, but mostly when travelling. As soon as the plane was off the ground, they had time to talk about their work. They would spark ideas off each other. In the course of discussion, they would sometimes identify new research subjects. For example, after a conference in Boston in the US, they gave themselves a weekend break at Cape Cod. *"We were in a gorgeous hotel with a sea view. Françoise and I spent hours walking on the beach among the seagulls, talking about the new subject that had caught our interest. It's a wonderful memory"*, he reminisced. That afternoon, they were talking about electrically charged polymers, "electrostatic spaghetti" which, because of electrical interactions, behave very differently from classical polymers. On their return, they enlarged on the weekend's ideas and drafted an article. They would also sometimes start this kind of scientific discussion at home. Pierre-Gilles would get out his pencil and describe his problem, scribbling on the first piece of kitchen towel to hand. He admired Françoise's critical mind, her *"knack for putting her finger on the problem"* and regretted the fact that many scientists wrongly attributed her results to him.

Polymers at the Beach

Pierre-Gilles did extensive research on the behaviour of polymers in the presence of a solid body, for example a wall. When they come in contact with such an obstacle (with which they have an affinity), the polymers try to cover it completely, by flattening themselves to it. At the same time, they follow their natural tendency to form coils. The theorist showed that the balance between these two effects depends on the concentration of the polymers. In small numbers, they adhere to the wall along their whole length, like holidaymakers stretched out on a deserted beach.[1] But just as some people have to sit as the beach fills up, in larger numbers the polymers only adhere to the wall by a few segments and form fragmentary coils. Finally, when the beach is full, the holidaymakers have to stand, and the polymers stick to the wall by a single segment, so the chains become entangled as in a semidilute solution. Pierre-Gilles established that the concentration profile of a polymer layer adsorbed on a wall follows a scaling law, which brings the size of the blob into play.

He developed the concept of the polymer brush to describe a somewhat similar situation, where the polymers are grafted to the wall, with his great friend Shlomo Alexander, during one of the latter's regular stays at the Collège in 1976. The two men had met at the Weizmann Institute in Israel in the 1960s and immediately became friends. They were as alike as twins, both very tall and slim, though Shlomo Alexander was almost skinny, with prominent cheekbones and hollow cheeks. They were both brilliant theorists, but Shlomo Alexander was more of a dreamer, with a tendency to absentmindedness, and did not have de Gennes' communication skills.

---

[1] J. C. Daniel and R. Audebert, "Small volumes and large surfaces: the world of the colloids", *La juste argile*, EDP sciences, 1995.

Together in 1976 they built a model using blobs which bears both their names (the Alexander-de Gennes approximation).

Oil Crisis

While continuing to work on polymer science (polymer adsorption on small particles in suspension (A11-6); plastic welding (A11-7); polymer flows, where he discovered a new "raw spaghetti-cooked spaghetti" type phase transition (A11-8), etc.), Pierre-Gilles was open to new topics. In October 1973, after the first oil crisis, he was contacted by engineers from the *Elf* Corporation in Lacq for help with one of their problems: they could not extract more than 15% of the oil in a deposit by direct pumping, since the rest remained trapped in the rock pores by capillary forces. The engineers could extract part of this by indirect pumping, i.e. by injecting water into a well to squeeze the oil into the neighbouring well, but obtained little more than an extra 20% by this technique (A11-9). They were therefore planning to use a surfactant-based fluid (containing molecules with a hydrophilic tail and a hydrophobic head) which, when injected into a well instead of water, they estimated would allow them to extract up to 50% or even 60% of the oil. The problem had all the ingredients to attract Pierre-Gilles. *"It was new to me... It immediately interested me"*, he admitted.

Added to a mixture of oil and water, the surfactant attaches to the interface between them and reduces the interfacial tension (the force needed to break the surface tension between the two liquids), like UN peacekeepers between two warning forces. However, there comes a moment when the molecules, instead of continuing to bind together at the interface, clump together to form micelles[2] which disperse into the water or oil, as if the peacekeepers were to go off for a drink rather than staying on patrol. The addition of a cosurfactant (e.g. an alcohol) destabilises these micelles, so that the surfactant can again mass at the interface (the peacekeepers return

---

[2] Micelles are balls whose surface is made up of the surfactant molecules, positioned with their tails inwards and their heads outwards, or vice versa.

to their duties). The interface then stretches further and further, to form a very long film between the oil and the water, to the point that the fluid becomes a dispersed medium of oil-in-water or water-in-oil, instead of being separated into visible water and oil phases. This dispersed medium can be ordered — when the film is organised into strips or stacked tubes — or disordered. In the latter case, it is called a microemulsion and takes the form of a sponge (when the quantities of water and oil are equivalent) or microdrops.

"Although microemulsions were already known, nothing very much was really understood about their properties, especially those associated with that huge interface between oil and water", explains Christiane Taupin, a physicist at the Collège de France lab. Pierre-Gilles fully understood its significance, showing that this interface governs the stability of microemulsions (A11-10). At the same time, he encouraged experimental work on microemulsions (at the Collège lab, but also elsewhere), and modelled the results. *"In a few years, we developed a microemulsion industry in France which has become one of the best in the world. It's another of the things I like to crow about."*

Ubiquitous Percolation

Pierre-Gilles de Gennes also managed to model the actual process of enhanced oil recovery, by unearthing some research he had done almost 20 years earlier. The injection of fluid into oil-bearing porous rocks reminded him of percolation in a coffee machine: the fluid gradually reaches different pockets in the porous rock and, at the precise moment when all the pockets connect, the oil spurts out, just as the coffee flows out of the percolator once it has passed through the grounds. So he described the process using the concept of percolation he had reinvented in 1959 (the concept had originally been introduced by an English mathematician in 1957)[3] and would apply it in the following years to phenomena as varied as gel formation (A11-11), conductive transition in microemulsions (A11-12),

---

[3] See Chapter 5.

magnetism in superconducting particles (A11-13) or suspension flows (A11-14). In the process, he showed the universality and power of the concept, as he explains so well in the popular science article published in the journal *La Recherche* in 1976. *"It's an amazing article: although it was written for a mass audience, it is still quoted in scientific publications today as one of the foundation articles of percolation theory — which is rare, probably unique. Pierre-Gilles de Gennes had this ability to formulate simple explanations of complex scientific phenomena that could interest both a mass audience and scientists"*, comments the physicist Michel Laguës.

Going Dutch

In 1978, he added a further distinction to his already considerable stockpile: he was invited to take up the honorary position of *Lorentz Professor* at Leiden, Holland's oldest university. From March onwards, therefore, he would catch a train to the Netherlands every week to give a class. *"It was a long journey, so I used it to work"*, he recalled. Before beginning his first class, tradition demanded that he sign his name on the back wall of the old lecture hall. Pierre-Gilles could pick out the signatures of illustrious predecessors — Felix Bloch, Nobel Prize in 1952, Lars Onsager, Nobel Prize for chemistry in 1968 — and also some of the teachers of his youth, like Rudolf Peierls and John van Vleck. *"It was intimidating"*, he admitted. His signature is still there on the wall. One of the perks was a superb apartment near the sea, which he rarely used. At most, he would have the odd dinner with physicist friends like Peter Mazur or Sybren de Groot and his wife Sylvia, a historian of Surinam, formerly Dutch Guyana, which he loved to hear about. *"I asked her for her impressions of my ancestor Jean-Baptiste de Gennes, an admiral at the time of Louis XIV, who set off to discover South America in 1695 and became governor of Guyana, and they were somewhat negative. Apparently, my ancestor did not govern the colony very well. He had tried to create an original industrial process by using watermills to press sugarcane, but they were swept away by the first flood."*[4]

---

[4] See the appendix on the history of his family at the end of the book.

Wetting

Pierre-Gilles moved naturally from the study of drops of oil trapped in rock to that of drops of water deposited on a surface, because the forces involved are similar. He began the study of wetting in 1983 and, of all his work on soft matter, it is this that gave him the most satisfaction, the sense that he had answered *"questions that had been left hanging for a long time"*, such as: a drop that shows an affinity for a medium tends to stretch and gains energy in the process, but where does this energy go? The weeks passed. Whichever way he looked at the problem, he could not find a solution. *"Finally, one day, I understood, thanks to observations dating from... the 1940s. In fact, as the drop advances it is preceded by a precursor film, like a carpet unrolling in front of it. The experiment that helped me to understand is very simple: you put a needle under a microscope, and deposit a drop of water nearby: as soon as the film reaches it, the needle falls. There's a great basic experiment performed with minimal apparatus! From this, we showed that all the energy gained through stretching was used in the production of the precursor film, and we understood why the film formed, what shape it was, etc. So we made great progress in the dynamics of this type of so-called "total" wetting"*, he concluded.

He also explained another strange phenomenon. A drop of water lies on a surface with a well-defined angle of rest. So far, no surprises. But when more water is added (using a syringe), the angle it forms with the surface is greater than the angle of rest, so that it only starts to stretch again from an angle greater than the angle of rest (subsequently reverting to that angle). Conversely, when water is sucked out of the drop, the angle it forms with the surface becomes smaller than the angle of rest, before the drop retracts. Pierre-Gilles demonstrated that this so-called hysteresis effect is caused by defects in the surface (asperities, chemical contaminations, etc.), which grasp and hold back the contact line (the edge of the drop). This line becomes deformed and, like an elastic band, accumulates energy. Beyond a certain degree of deformation, it is released and relaxes.

The run-up to Christmas 1983. Pierre-Gilles is in the process of constructing this theory at the Collège with his doctoral student

Jean-François Joanny. On December 22, he erupts into his student's office. Joanny looks at his boss in astonishment: what could have made him lose his cool? *"Pomeau* (Author's note: a theorist at the École normale supérieure) *has had the same idea as us. Quick, we have to write the article! Give me your notes."* The student collects his sheets of calculations and hands them to Pierre-Gilles, who puts them under his arm and shuts himself in his office. *"He came back every two hours with four or five pages for me to proofread. In all, wrote some thirty pages in the afternoon"*, marvels Jean-François Joanny. Around 7 p.m., visibly tired, Pierre-Gilles enters his student's office one last time. *"I have to go. You'll have to finish the article over the holidays and get the secretary to type it up on the first Monday back. Make sure that Pomeau gets a copy."* On Monday, January 2, 1984, as agreed, Jean-François Joanny left a copy of the article in Yves Pomeau's empty office. As he was leaving, he met one of the physicist's colleagues, who told him that Yves Pomeau had come back from his discussion with Pierre-Gilles de Gennes equally excited, also emphasising the need for speed. Ultimately, Pierre-Gilles could produce at a rate that was hard to compete with, and Yves Pomeau's article would be published much later.

In broad terms, Pierre-Gilles revealed the significance of long-range so-called van der Waals forces (responsible for the cohesion of molecules in liquids) in small drops (A11-15). His results were supported by an experiment which shows that in a situation of total wetting, a drop does not stretch along its full length to form a carpet of molecules, as one might imagine, but forms a thicker "pancake" because of these van der Waals forces. *"The dynamics are amusing. Initially, it's like an Aztec pyramid, with superimposed layers of equal thickness (one or two nanometres). I calculated that everything happens near the edges and nothing at the centre. It is the edge that moves out towards the layer below, until eventually there is just one layer"*, he explained.

In a period of three years, Pierre-Gilles contributed more to the physics of wetting than had been achieved in fluid mechanics in 15 years! Once again, it was his approach that made the difference. *"You mustn't get obsessed with precise calculations. In wetting dynamics, very good researchers, for example Skip Scriven at Minnesota University, have calculated the dynamics of drop movement in certain conditions: it's impeccable, but you can't deduce anything from it. Whereas after my calculations, a very simple formulation emerged from which you can see what happens — it's crude, but at least you can learn something from it"*, he argued. Not everyone rejoiced at the breakthrough. *"Fluid mechanics specialists are often very strict applied mathematicians. Unfortunately, I have always clashed with them. They tell me, either at conferences or in letters: "It's vague, it's hard to see where you're going..." It doesn't seem to be a generational issue, because it's been going on for years. The culture in fluid mechanics seems to stay pretty much unchanged"*, he commented.

Soap Bubbles

After water droplets, the theorist turned his attention to soap bubbles. *"Some bubbles are perfectly simple and well-behaved; they show magnificent colour effects. There is no major disturbance or agitation. Conversely, you can see with soap, simply by washing your hands, that certain bubbles show frenetic activity"*, he noted at the start of his lecture on "Bubbles, Foams and Other Fragile Objects". He explored this activity at the surface of bubbles by studying the ageing of a film of soap obtained by pulling a stick out of a bath of soapy water (the agitation is linked with the temporary nature of the bubble). The film is unstable and ages by thinning from the inside, where gravity causes the water to flow between the soap surfaces (A11-16). Pierre-Gilles had assumed that the flow remained constant during ageing. *"I gave the subject to Élie Raphaël and Achod Aradian, who found different scaling laws from those I had in mind. They showed that they were no longer in the little league"*, he rejoiced.

In general, the theorist dealt well with his mistakes, even finding them easier to talk about than his successes. For example, when he realised that his article on the stabilisation of foams by polymers, published in the *Proceedings of the Academy of Sciences* in 1979 was wrong, he made amends by publishing a second article in the same journal, which attracted mockery from André Guinier. *"You write one article. Then you write another one in which you say the opposite. In that way, you end up with two articles"*, he jeered.

Dewetting

One evening in February 1987 at Les Houches, after classes at a winter school on surfaces, in the pleasant little chalet where they were staying, Pierre-Gilles shared his latest thoughts on the thinning of soap films with Françoise. *"Ultimately, isn't a film of soap like a film of water? Couldn't your description be transposed to the more general case of films of water on a surface?"*, was her simple suggestion. He thought for a while, considering the forces involved in both cases. *"It's true, by adjusting coefficients it should be possible to draw a parallel between the draining of a film of soap and the drying of a film of water."* He smiled in admiration of Françoise's stroke of genius. *"Her idea led to years of research on dewetting"*, he recalled. Clever as he was in finding answers, Pierre-Gilles de Gennes was also very good at seizing on the right questions.

He thus showed that the two phenomena (the draining of a film and dewetting) obey similar laws. When a film of water is deposited on a hydrophobic surface, dry areas appear spontaneously (usually beginning at defects on the surface) and then expand. They have a rim which sucks up the water as they widen. The water at the rim then flows forwards, between the surface and the interface with the air, just as the water in a film of soap flows between two surfactant walls. The theorist also showed that the dry areas grow at a constant rate. *"The theoretical study took six months, but would take four years to prove experimentally"*, he specified. He soon moved on from these water-related issues. *"I had a lot of fun with capillarity and wetting. My range of interests proved very useful, enabling me to solve problems at 50 more quickly than at 30."* In 1992, he published a textbook called *Capillarity and Wetting Phenomena: Drops, Bubbles, Pearls, Waves*. *"Françoise and I had been planning it for some time"*, recalls physicist and co-author David Quéré (he did his thesis at the Collège). *"The idea was in the air, but neither of us had really made any progress. One day, Pierre-Gilles came along and said that we really needed to write the book, otherwise someone would get in first, and suggested that we should do it together.*

What a godsend! The book was written in a few months. On Fridays I would find notes on my desk: "Don't forget to give me your next chapter on Monday for proofreading." Fortunately, I had no other plans those weekends..."

Mr. Superglue

While Pierre-Gilles felt that he had *"made a contribution"* to the field of wetting and dewetting, he had reservations about the usefulness of his work on adhesion, despite the fact that he would be introduced — somewhat absurdly — as *"Mr Glue"* or *"Mr Superglue"* after the award of his Nobel Prize. He started on the subject in 1989, but quickly realised that the question he wanted answered — *" Why do adhesives stick?"* — was easier to state than to answer. Adhesion is a complex problem. *"There is not just one glue, but thousands of different glues"*, he would repeat. So there was not just one answer to the question... The first thing is that a glue is a polymer joint that holds two solids together. To test its quality, one tries to separate the two solids, pulling harder and harder until the joint breaks, which determines the breaking threshold and, therefore, the quality of the glue. However, this breaking threshold depends on the way the glue has been applied! A real puzzle. Pierre-Gilles then established the separation energy of the two solids and showed that it is more relevant than the breaking threshold in assessing the quality of the glue. He specified the breaking mechanism of a joint made with one variety of glue (elastomer), highlighting the importance of the stretching of the polymer chains (A11-17). *"But I never understood how the other glues, for example superglues, worked"*, he confessed. *"And then, I realised that I wasn't capable of giving a class on adhesion that was guided by ideas rather than recipes. For me, that's an infallible sign: adhesion is not really a success."* His only consolation would be that, by explaining the fracture mechanism from polymer scale phenomena, he had built a bridge between the world of chemistry and the world of fluid mechanics.

Jack of All Trades

Because of the diversity of the subjects he tackled, as evidenced by the pages above, Pierre-Gilles de Gennes has sometimes been described as a jack of all trades. However, these subjects are not as far apart as they might seem. In fact, the theorist followed a guiding thread, genuine though not apparent to everyone, which led naturally from one subject to another, for example from soap bubbles to foams, etc. Before starting on a new path, he would simply make sure that his approach would bear fruit. No matter if the field was unfamiliar... He was ready to *"acquire the necessary equipment en route"*, as he would say, provided that the game was worth the candle. Over the years, therefore, he assimilated a massive range of fields — hydrodynamics, surfactant chemistry, fracture mechanics, etc.: he was a living encyclopaedia. He was also ready to explore questions that emerged from the different subjects: for example, what happens to liquid crystal polymers, or how does dewetting take place in a liquid film when the surface is covered with rubber? With this wide range of subjects available to him, he was never short of ideas. When he got stuck on one theme, he would work on another, before reverting to the previous one, sometimes years later. So he always had several irons in the fire.

Whether with polymers, wetting, adhesion, etc., he never started his research from scratch, but always relied on previous work, *"usually fragmentary and incomplete"*, stresses his colleagues Liliane Léger. In this sense, therefore, he was never a pioneer, even though he might be seen as one. The fact is that he tackled these subjects from such a different perspective so differently that he gave them a new shine, as if he had restored some old objects found in an attic, making them look new again. He was not interested in old things, but in day-to-day objects that theorists had so far usually ignored. The CD-ROM on soft matter that he made in 1997 with Madeleine Veyssié exemplifies this: he invites us to pick out an object in a house, for example a laptop computer or a tube of glue, and to ask ourselves — how does the LCD screen work? Why does the glue stick? etc.

By contrast, all these objects — polymers, interfaces, emulsions, etc. — were of key interest to industrial firms, both for their applications (detergents, coatings, paints, plastics, cosmetics, etc.) and for their processes (extrusion, lamination, sintering, etc.). The mid-1970s marked the start of a close collaboration between Pierre-Gilles and industry.

Until then, he had certainly been interested in applications, in contrast with most physicists of the time, who in those post-1968 years looked down on any engineering-related questions, and refused to deal with an industry seen as being *"in cahoots with capitalism"*. Nonetheless, his links were usually confined to discussions on fundamental questions with theorists from companies like *General Electric, Exxon* or *Bull*. After his study on enhanced oil recovery, he started to listen to industrialists in the hope of finding solutions for their problems.

He had everything they wanted: not only did his research topics match their needs, but his pragmatic and results-oriented approach to problems facilitated relations. *"When you're working with industry, you're not looking for exact solutions. You could always spend several weeks inputting all the equations defining a problem into a great big computer and eventually get a behaviour law for a specific case. But you wouldn't know what to say if you were asked: "What happens if we change this variable?", because you wouldn't know which parameters were important. By contrast, with the scaling laws approach, you can discuss the important parameters with the workshop foreman"*, he explained. So he received more and more approaches from industry. He always answered and was happy to visit their labs, for example *DuPont de Nemours, Johnson & Johnson* — *"They were doing research on insecticides and breeding a frightening number of nasty critters. The lab was depressurised, so that they couldn't get out, but it was still impressive"*, he recalled — *Kodak, IBM, Philips, Shell, LVMH, Nestlé, Sanofi...*

These visits sometimes caught him by surprise. In April 1998, engineers at Unilever's research centre near Liverpool in the UK told him that the soap manufactured in their factory varied in quality according to the... lunar cycle. Pierre-Gilles burst out laughing, but it was not a joke. No one could understand how the moon might affect the production process, but that was the reality. In fact, he learned a few weeks later that the water used in the factory was pumped from wells a few miles from the sea and its salt content varied with the tides, which is why the quality of the soaps depended on the phases of the moon!

Not all Pierre-Gilles' relations with industry led to scientific work, but some proved fruitful. In 1982, for example, at the end of a visit to one of the *Allied Chemicals* labs on the USA's east coast, his hosts told him about a *"highly technical"* problem. *"When we mix powdered polymers into the solvent, they dissolve too slowly. We aren't able to speed up the process, even by grinding the powders more finely"*, he explained. Pierre-Gilles responded: *"It's simply a problem of solvent diffusion."* The American answered politely that the problem was unfortunately more complex than that. On the flight back to Paris, the French physicist worried at the question until he found a solution. Roughly speaking, the powder dissolves slowly because the deformation of the mesh formed by the polymers inhibits the movement of the solvent. Pierre-Gilles provided a solution by establishing the optimum size of the particles in the powder, but the discussions were beneficial in both directions, since this *"highly technical"* problem would lead to years of basic research. Similarly, the problems of mottling — a cloudy printout that occurs when the water films used to protect the colours previously deposited on the paper have not been completely evacuated — that Pierre-Gilles encountered when visiting the Arjo Wiggins paper factories, launched him on a new line of research. *"Their problem related to familiar dewetting issues, but it was different in this case because the water was partially penetrating into the paper. We were confronted with a new question, so it was stimulating. This problem gave rise to a series of articles on dewetting in a porous medium, etc."*, he explained.

In practice, the collaboration developed not only through visits, but also through seminars and PhD funding. When Christiane Taupin became head of department at *Atochem* (at the time the chemicals arm of *Elf*) in 1986, she did not lose contact with the lab, since the company was financing several doctorates there. *"As far as Atochem's bosses were concerned, the aim was that I should maintain relations with the lab, and... with Pierre-Gilles de Gennes"*, she reports. The Collège de France laboratory was not short of contracts, and the research that went on there owed much to industry.

Through working with engineers, Pierre-Gilles became more aware of productivity and cost constraints, which had repercussions on his research. He would sometimes lose interest in research that, in his opinion, offered no industrial benefits. In 1986, for example, he was initially enthusiastic about dendrimers, tree-like molecules with repeated branchings, which can be synthesised to order. He thought that they might be useful as vehicles for medical drugs. But he changed his mind when he realised that they were too sophisticated, therefore too costly to produce, for biomedical applications. *"I had done a calculation on the organisation of these molecules into cauliflowers (but I think it was wrong). From an industrial perspective, it is more advantageous to synthesise molecules statistically, introducing the components randomly, rather than to order"*, he concluded.

Capitalist Dictatorship

Industrial firms courted him as a consultant, to sit on their scientific committees or their boards of directors, lucrative positions which he either accepted or declined. The honeymoon lasted until the 2000s, when Pierre-Gilles realised that the big corporations were significantly shifting their research policies. *"In the first half of the 20th century,* DuPont de Nemours *invented nylon,* General Electric *silicon and in 1948* Bell Telephone *invented the transistor, the starting point for the electronic civilisation we live in today. Alongside their industrial activity, these firms worked within a long-term perspective. Alas, this attitude is fast disappearing"*, he complained. He went on to question the rise of shareholder power in the 1980s. *"The pension funds vetoed funding for any research that would not increase turnover and profits [within three years]."* And turned his ire on the investment funds, *"those predators"* which put their weight behind mergers *"that have killed entire swathes of future-looking research."* He gave the example of the closure of an ultramodern agronomic research centre near Lyon, as a result of the merger of *Hoechst* and *Rhône-Poulenc* in 1999.[5] *"What I observed is the (dramatic) decline of long-term research in the big industrial conglomerates. And its impact on state bodies which, these days, want to convey the same image of immediate profitability"*, he wrote in 2000. A trend that is still very much in evidence.

---

[5] Philippe Genet, "France is experiencing an unprecedented brain drain", *Capital*, December 2003.

His industrial involvement led to even more travel. He was never in France more than three months at a time. Travel punctuated his life, with its inevitable surprises: a typhoon in Japan — *"I was on a train to Okayama when the sky turned the colour of judgement day. A typhoon was on its way. Pylons were falling. We had to be evacuated to a village"*, he recalled; a coup d'état in Pakistan in 1977 — *"After a summer school in the Himalayas, we had returned to the capital and had attended a reception at the US Embassy, with all the VIPs. I went to bed fairly early, around midnight. At 3 a.m., there was a coup. Prime Minister Zulfikar Ali Bhutto was overthrown and the generals took power. The Pakistanais with us were nervous, then things calmed down after a few hours."* He had a hundred stories like this to tell.

He also travelled during his vacations. In the summer of 1991, after a conference in Corfu, he went to stay in Portugal with Françoise and their children. *"We stayed in a sort of youth hostel, in small cabins, 100 km north of Lisbon. Claire and her girlfriend were living the life of Riley, probably more than we suspected, while I looked after Olivier: I was overseeing his school assignments, but he wasn't very interested. He was already somewhat anti-school. In the evenings, we all ate together in a huge dining room. It was fun, except that they invariably served us sautéed potatoes"*, he reminisced. In August, he went with Annie to stay with Louis Blanchard, in the north of the Luberon region. *"Then we stayed with other friends, in the Cévennes with Henri Benoît, in the Aubrac and in the Massif Central. A beautiful summer..."*, he recalled. It was also his last summer of anonymity.

Chapter 12

School Director

Let us go back a few years to talk about a major event: in September 1976, Pierre-Gilles de Gennes became Director of Paris's prestigious ESPCI — Higher School of Industrial Physics and Chemistry (nicknamed PC). Created in 1882 at the heart of the city's 5th arrondissement, this School was proud of having welcomed within its walls great names like Pierre and Marie Curie, Paul Langevin, Frédéric Joliot, etc. The appointment of Pierre-Gilles made waves, because for the first time in the school's history it broke the unwritten, but hitherto strict law that the director should be a former pupil of the school, which Pierre-Gilles, as a graduate of the ENS, was not. What had happened for this almost century-old tradition to be overturned? In fact, the appointment arose from an imbroglio of intrigue within the ESPCI.

A year before, the obvious incumbent for the directorship of the school was a professor whom the then director, the charismatic and paternalistic Georges Champetier, had taken under his wing and appointed director of studies.[1] However, this professor's impetuous temperament had turned some staff against him, to the

---

[1] The director of studies was responsible for all matters of pupil selection and teaching at the school.

point that they decided to put forward a replacement candidate. Easier said than done. They soon realised that none of the teachers at the school would fit the bill, and therefore decided to look for an outside figure, who would necessarily have to be of greater scientific stature than the director of studies. Pierre-Gilles de Gennes seemed an obvious choice. Not only was his scientific stature beyond dispute, but he was also interested in both physics and chemistry, and — the icing on the cake — in industrial problems. He was the perfect man for the job! Too bad if he wasn't a former pupil. But he still had to accept... Why on earth should he get involved? He already had quite enough to do. Lucien Monnerie, who had worked with him on polymers, was given the task of making the case. He went to the Collège de France, but felt so sceptical about his chances of success, that he finished his spiel with the words: *"Ultimately, I don't see why you should agree to come."* Pierre-Gilles kept a poker face. He was not very surprised by the proposal, because Jacques Badoz, a professor of optics at the school, had mentioned in an aside on a greetings card in January that Georges Champetier's job would soon be up for grabs. The theorist had read between the lines. *"I'll think about it. I like the school, a nice size for a grande école"*, he told Lucien Monnerie.

Pierre-Gilles was actually familiar with the ESPCI, having taught there until 1968. As he said, it was a small school, with 250 students and a hundred or so teachers and researchers. *"After making enquiries, in particular of Jean Brossel at the École Normale Supérieure, I had reached the conclusion that there were a few labs worth their salt amongst the dozen or so at the school. That encouraged me"*, explained the physicist. A few weeks later, he saw Lucien Monnerie again.

— *"I am prepared to head the school for a maximum of five years. It wouldn't be a bad thing for me to get out of my ivory tower at the Collège and develop my connections with the world of industry".*
— *"If you accept, you will have to commit for at least ten years: it's the minimum time needed to undertake in-depth reforms".*
— *"Well, you know, men in my family die young. We'll see..."*

When the director of studies was informed of Pierre-Gilles' candidacy, he withdrew his own, later admitting: *"For anyone but you, I would never have backed out."* The way was now clear. In 1976, aged 44, Pierre-Gilles officially became director of the school. It should be noted that he attached draconian conditions to his acceptance. *"I will not spend more than two days a week on the job. This means that I need strong backup"*, he demanded. A tailor-made structure was immediately instituted, around a trio consisting of Lucien Monnerie as director of studies, Jacques Badoz as scientific director[2] and Jean Léoni as general secretary, who managed the life of the school, only keeping Pierre-Gilles informed, at his request, of major events. He was not interested in hearing about details or administrative issues. *"At the management meetings, which took place every week, as soon as the subject turned to management issues, you could see that his attention wandered. He would even use the time to go through his mail. But at the slightest digression, he would bring us to order: "we're getting off the point", he would say. So he was listening despite appearances. He was happy to delegate and placed complete confidence in us"*, recounts Jean Léoni. Indeed, he let them run a number of major projects without getting involved.[3] By relying on this trio, he was able to manage the school while still devoting most of his time to research.[4]

However, it demanded a very strict timetable. He would arrive at the school at 8:30 and spend the morning in his huge office. Like the one at the Collège de France, he had arranged it to his taste, installing coloured panels for use as writing boards and scattering family mementos around the place, a Roman marble inherited from

---

[2] The scientific director was responsible for research done in the school's laboratories.

[3] For example, he gave carte blanche to Jean Léoni to create a student residence, and to Jacques Lewiner, scientific director after Jacques Badoz, to set up an association to develop start-ups in the school. Pierre-Gilles was delighted to see the school *"return to its tradition of innovation and patent acquisition."*

[4] He was so conscious of the strong support he received at ESPCI that he would refuse the offer of the directorship of the École Normale Supérieure a few years later, convinced that he would not find people at the ENS as devoted as at ESPCI.

his father, a collage by his wife and works by his sculptor daughter Marie-Christine, all creating the atmosphere of a comfortable living room. He would leave the school a little before noon, walk the 10 minutes to the Collège de France and then return around 5 p.m., before retiring at the end of the day to the service apartment above the offices. Annie would not join him till much later, when the *Boudin sauvage* stopped serving, leaving the van with the restaurant logo in the school courtyard.

Experimental School

In accepting the directorship of the school, Pierre-Gilles had seized a unique opportunity to put into practice his ideas on teaching and research, using the establishment as a sort of real-life laboratory. However, before beginning any major reforms, there was a small problem to resolve. *"Almost as soon as I arrived, I was told that there was a lecturer who never turned up at the school. Up to that point, everyone had turned a blind eye. I managed to get him into my office. He was extraordinary, with a white beard almost down to the ground and the look of a prophet. He spent his day doing nothing, but would leave home every morning saying he was going to work. We had to advise him to seek medical help"*, he recalled. He quickly put a stop to this situation, which had been allowed to go on too long, setting the tone for his term of office.

The new director began by tackling teaching reform. *"In Cambridge, I found an interesting system"*, he announced. *"Students work under a supervisor or tutor, who is familiar with their strengths and weaknesses and provides the necessary support."* The project was launched. As director of studies, Lucien Monnerie spent a year researching the question, *"with typical conscientiousness"*, noted Pierre-Gilles de Gennes, travelling to Cambridge, Oxford and London to explore the ins and outs of the English university systems and establish a tutorial system appropriate to the school. He spent a second year presenting it to the teachers. *"The aim of the tutorial system is that students should be able to learn for themselves and acquire the basics of each discipline, assimilating the significant orders of magnitude and the idea that the same laws apply in different domains"*, he explained. *"Lectures will now focus on essential concepts, whilst the examples will be covered in tuto-rial sessions."*

The teaching curriculum was to be entirely remodelled. The number of hours of lectures was cut by a third; TD (supervised lessons) were eradicated; even the examinations now offered a choice of different subjects. A revolution was underway! Some teachers came on board immediately, but others dragged their feet, reluctant to change their habits. *"It was the same thing every time: when the number of lecture hours was reduced, they said that it was unacceptable to encroach on their discipline. It was as if they took it as a personal insult"*, relates one teacher. Tensions ran so high that one professor insisted that it should be recorded in the minutes of the board meeting of October 28, 1980 that professors *"fear that [the reduction in the number of lectures] would result in a diminution in the knowledge acquired by the students and in the standard of the school."*

Nevertheless, the tutorial system was introduced at the beginning of the academic year in 1980. It was a first in France's educational system. It was relatively costly, with the salaries of tutors and the purchase of hundreds of books, etc. For the occasion, it was Pierre-Gilles rather than the general secretary who welcomed the new students. He entered the lecture hall, raised his hand in greeting and sat down on a corner of the table, smiling at the pupils. Even before he spoke, they had already grasped that his directorial style was unusual. Then he stood up and, pacing around the podium, explained that he had two expectations of them. The first was that they should be able to calculate orders of magnitude. *"When he met his students, Enrico Fermi used to ask them to estimate the number of piano tuners in New York . How would you go about it? The city has a population of 10 million. On the assumption that one family in 30 has a piano, there is approximately one piano per 100 inhabitants, i.e. 100,000 pianos. Assuming that a piano has to be tuned every three years, 100 have to be tuned every day, and further assuming that a piano tuner tunes an average of two pianos a day, there are around fifty piano tuners in New-York. That is how you should reason"*, he explained. He was not alone in recommending the use of orders of magnitude, but he made it a basic educational rule, a constant feature of his work. The second quality required was the ability to observe. *"When walking on the shores of the Baltic Sea, the Swedish oceanographer Vagn Walfrid Ekman noticed that blocks of ice driven by the wind blowing from the west did not travel precisely in an easterly direction. On reflection, he concluded from this observation that the rotation of the Earth was responsible for this drift. He arrived at this discovery because he noticed an abnormality in the direction of travel of the blocks of ice. That is how you should observe"*, he concluded. For an hour, the students listened openmouthed. These words contradicted all the homilies they had previously heard, and would mark some of them than ever. *"That speech counted for me at least as much as everything I learnt at school"*, asserts David Quéré, a former pupil at the school who worked alongside Pierre-Gilles. Every year until his retirement, Pierre-Gilles would personally welcome the new students with similar sentiments.

In general, he hardly saw them after that, leaving it up to his director of studies to deal with the students' academic choices and concerns, and to Jean Léoni, the general secretary, to track the ups and downs of their lives (depression, anorexia, dropout...), knowing him to be close enough to the pupils for them to share confidences with him. *"Jean Léoni looked after them like a mother"*, he recalled admiringly. If they had a problem, Léoni was the one they went to, not the director.

To begin with, the tutorial system lived up to expectations. There were plenty of volunteer tutors, and the students played the game. Over the years, however, the system gradually lost sight of its objectives, with some undermotivated tutors turning the sessions into ordinary supervised lessons or remedial classes. On the student side, things were not much brighter: some pupils would swap tutorial subjects and cram for exams. It may be that this loss of direction was at least in part attributable to the idealism of Pierre-Gilles, who imagined that the students were like him at their age, with an unusual thirst for knowledge and appetite for work. He nonetheless persevered in trying to make them more independent and in promulgating fewer hours of lectures, but given the hostility of certain teachers, he would eventually become less interested in the teaching at the school than in research. He undertook drastic reforms and committed himself to carrying them through, as if he felt a personal stake in the excellence of the institution.

Hand-To-Hand Combat

On his arrival at the school, his assessment was stark. *"Research here lives off the legacy of past glories, from the time of Pierre and Marie Curie."* The message was clear: he wanted a clean sweep of the ageing labs (sparing the few that enjoyed a good reputation). A long struggle ensued. Not all the researchers greeted him as the messiah, some indeed taking a dim view of his arrival. His declarations generated anxiety: the researchers wondered what was in store for them. It would not be long before they found out. From the start, Pierre-Gilles set the tone.

As ever, he began by listening to people's views then made his decision, with no further argument. The first measures related to research funds. Having long been distributed at the director's say-so, since 1968 they had been allocated by a research committee made up of duly elected professors at the school, which in fact essentially redistributed the same amounts year after year. *"The school needs a scientific policy, [which] runs counter in its principle to the notion of equal allocation. Funding variations sufficiently sharp to provide incentives need to be [introduced]"*, he asserted at the meeting of April 26, 1979. And he described the policy that he planned to develop. *"Applied research and fundamental research are equally important if they are innovative and original. In recent years, the climate in France has been to prioritise research for short-term applications; we should not follow this trend (...). [Our] policy cannot be determined by the policies of outside bodies (CNRS, industrial concerns, etc.)"*, he explained. He went on to denounce what he called *"the persistence effect"* (the fact that *"a scientific field can lose its vitality without (...) any apparent diminution in its activity"*) and *"the acceptance of contracts less and less connected with any original research effort"*, in other words providing services for companies, which had become the sole activity of certain laboratories. To stimulate innovation, therefore, he installed a new system of "merit-based" allocation, dedicating part of the funds to research projects (not labs), and set up a laboratory assessment committee.

These measures caused uproar. The professors were not used to people poking their noses into their affairs. Above all, they opposed the evaluation committee, because it was headed by Pierre-Gilles and it was he who wrote the reports. They objected that its impartiality was not guaranteed, because the sole purpose of the committee was to promote his objectives. *"You can't be judge and lawmaker at the same time"*, they complained. Despite the protests, he continued to head the committee and evaluated the labs himself, issuing good and bad grades, with the associated financial repercussions. He did not go at it with kid gloves and instructed certain professors to change subject immediately, describing their research as *"completely archaic"*. The climate became poisonous, but Pierre-Gilles held his course.

"It is very difficult to assess research. It is often said that foreign researchers should be used for the purpose, but they are often careful not to seem ill-mannered and, in the end, make somewhat "soft" assessments for fear of being too rude about a laboratory or a researcher. Internal evaluation systems can be useful, although there are different considerations such as the relations between researchers, which can affect the conclusions of the report. What we need are evaluation systems based on small groups of scientists, capable of being incisive and robust, like the ones Allègre set up in the 1990s, but that requires top-level scientists, and because the evaluations take a lot of time, they would have to be paid...", he regretted.

The evaluation committee was not the new director's only controversial measure. Observing that some professors were neglecting their laboratories, too busy with their different functions in industry and academia, Pierre-Gilles began a clampdown on job accumulation that also raised hackles. However, the animosity reached boiling point when he attacked the oligarchy in the school, where the laboratories essentially ran like "fiefdoms", headed by an all-powerful professor who laid down the law, generally recruiting deserving students to serve as his subjects, with the result that the school operated almost in a vacuum. Pierre-Gilles wanted to put an end to this *"inbreeding"*, repeating that *"the school needs new blood"*. As a consequence, just as he brought new teachers and outside tutors into the school, he encouraged the recruitment of researchers from outside labs. *"He knew vast numbers of top-level researchers from all over and would get enthusiastic about this or that person who had achieved such and such. His dream was to get them to come to the school"*, recalls Jean Léoni.

Despotic Reputation

Between barons deprived of their absolute power and researchers annoyed at not being in his good books, Pierre-Gilles made enemies at the school. It has to be said that he did nothing to make himself popular. For example, he took decisions without listening to people's grievances for as long as diplomacy might require — when he listened at all — and gained a reputation for authoritarianism. When he arrived at the school, he seemed distant, striding around without making eye contact. *"If I start to say hello, I'll never stop"*, he explained to a physicist friend who queried his attitude. Some complained: *"It's impossible to talk to him. And when we get an appointment, everything has to be sorted out in 15 minutes."* He quickly established a reputation as a despot.

Whether or not this is true, he did not have much room for manoeuvre. In reality, he could not fire a research professor and could only hire a trickle of new staff. So he had to wait for one research professor to go in order to appoint another. Nevertheless, over the years he succeeded in injecting *"new blood"*, his constant leitmotiv, and in creating new labs. For example, following the retirement of its director in 1977, he revitalised the mechanics lab by turning it into a cutting-edge hydrodynamics laboratory, appointing the loyal Étienne Guyon, one of his musketeers at Orsay, to run it. He also created a theoretical physical chemistry group in 1987, headed by Jacques Prost, a young and promising liquid-crystals specialist he had spotted in Bordeaux. A chair of mathematics became a chair for foreign guest professors, who would spend a few months at the school (the Joliot-Curie chair). Other labs came into being: "Waves and Acoustics", "Colloids and Nanostructures", "Soft Matter and Chemistry" and "Microfluidics, MEMs and Microstructures".

Ultimately, Pierre-Gilles was happy with the development of the labs, *"not in their size, but in quality and research subjects"*.

However, not everything he touched turned to gold. For example, he had intended to start research on suspensions by placing Pawel Pieranski, an experimentalist he had worked with at Orsay, at the head of the "director's laboratory" (a lab that the school's director could use as he wanted). Unfortunately, the physicist felt uncomfortable with the fiefdoms, and returned to Orsay after two years. He also had high hopes for a brilliant young researcher whom he had recruited on recommendation, but who gradually let his lab go to the dogs, forcing Pierre-Gilles to remove him from his job.

It was one of the toughest sanctions he applied. Because *"His Majesty"*, as he was now nicknamed, was usually quite merciful. A few months after his arrival, for example, on discovering that a professor was taking advantage of the contracts he had with various companies to grant himself certain privileges, he took no action. *"I was stunned — and I had to undersign the contracts — but I had no administrative power to oppose them* (Author's note: there was nothing illegal about these contracts) *"*, he explained. Nor did he punish the female employee who had thrown away some of the school archives, and not just any archives: they notably included letters from Albert Einstein to Paul Langevin. *"We had to make room. Where would the workmen have put the equipment otherwise?"*, was her explanation. Pierre-Gilles was upset, but took the matter no further. *"The damage was already done."*

He hated nothing more than conflict, though his position often put him in the role of arbitrator. For example, he had to intervene to prevent one of the school's toxicologists making a complaint against... her own lab. *"She used to keep her toxins carefully in a cupboard. One-day, her boss had everything thrown in the garbage: she lost all her research material. It very nearly went to court"*, recalled Pierre-Gilles de Gennes. He also had to intervene when one chemist turned up at the school with a knife in his hand, ready to butcher a colleague. *"I had several cases of that kind. There was also an occasion when I had to physically separate two researchers. One of them was the ant, very good at raising funds, the other the grasshopper, let's say more of a dreamer. Initially, the grasshopper was looked after by the ant and was very well off. After a few years, it went sour and they came to blows"*, he recalled sorrowfully. Although these matters irritated him enormously, he usually manage to keep his cool and resolve them calmly, even skilfully.

Incredible Discovery

One morning in September 1986, Pierre-Gilles was leafing through the contents of *Zeitschrift für Physik B Condensed Matter* in his office at the school, when a heading came to his attention. He immediately went to the article, and couldn't believe what he was reading. *"It's against all the odds"*, he exclaimed. The article described the discovery of a material, a copper oxide, that was superconducting at 38 K (−235°C).[5] He had been convinced that the critical temperature of superconductors would never exceed 25 K, as other theorists had demonstrated. *"Fortunately, I never published anything on it."* Then he took the measure of the fantastic scientific challenge that this opened up, because since the superconductivity of these copper oxides could not be explained by the usual theory (BCS), it could only be assumed that this was a new form of superconductivity governed by an unknown mechanism. Pierre-Gilles had an idea. These superconducting copper oxides in fact reminded him of the mixed valence manganese oxides that he had worked on at Berkeley in 1960, which had a similar structure.[6] He got out his old article and tried to explain the superconductivity of the new superconducting oxides through the mechanisms he had studied in 1960 (A12-1). However, he found experimental results that seemed to contradict his idea, as he explained in 1987 to Julien Bok, a specialist in solid-state physics at the ENS and a former fellow conscript (on the *Richelieu*, in Brest). *"Publish anyway. Your approach is interesting"*, the

---

[5] One of the authors was Alex Müller, whom Pierre-Gilles had known well since the 1960s. The two men had met at the IBM research labs in Zurich, where Alex Müller worked and Pierre-Gilles was a consultant. See Chapter 8.
[6] See Chapter 6.

latter replied. Pierre-Gilles followed his advice and published, but without conviction.

During this time, other copper oxides were discovered which become superconductors at even higher temperatures, first 93 K, then 110 K. They sparked unprecedented excitement in the research community: thousands threw themselves into the fray, some to capture the temperature record, dreaming of a room temperature superconductor, others to discover a new theory of superconductivity. The worldwide competition was feverish. At the same time, in France, a polemic had broken out in the scientific microcosm. It was realised that these copper oxides had in fact actually been synthesised in a laboratory in Caen, but no one had taken the trouble to test their possible superconductivity. Some claimed that if Pierre-Gilles had not dropped out of superconductors 20 years earlier, France would not have missed out on this discovery. The accusation was unfair, because other great theorists had stopped their research at the same time as the Frenchman: the field had been well and truly ploughed at the time (clearly, since it was another 20 years before anything new happened). In any case, he cared little about what was being said. In 1987, *"horrified by the number of physicists working on the subject and unable to see himself contributing anything whatsoever"*, Pierre-Gilles dropped the idea of continuing his research on high-temperature superconductivity. He kept informed of the dozens of models that were emerging, but none of them was convincing, and the mystery subsists to this day. Nevertheless, in 2007, he would return to the subject, discovering a mechanism that might explain everything...[7]

Although he withdrew from superconductor research, he encouraged it at ESPCI, creating a new solid-state physics lab. He appointed Julien Bok to head it, asking him to pool all the work begun by researchers at the school, caught up in the same fever as physicists all around the world. A few months later, one of them, Michel Laguës, spotted a conduction anomaly in layered bismuth-based systems. It looked as if they became superconducting at room

---

[7] See Chapter 15.

temperature. Some colleagues reacted to this astonishing result with enthusiasm, others with scepticism. The excitement was frenetic. Pierre-Gilles was dubious, wondering whether it was a new phenomenon or an experimental artefact: he called everyone into his office. Michel Laguës was well known to him, having spent two years in his lab at the Collège de France. After discussion, he reached his decision: *"Publish your result. If it is confirmed, you will be first. Otherwise, too bad."* The article came out in the prestigious journal *Science* and referred to superconductivity as a *"possible explanation"*. Nonetheless, it triggered a media storm — and lively controversy, as no one was able to reproduce the results. Sure of his ground, Michel Laguës continued his research, with the encouragement of the scientific director, Jacques Lewiner, who raised venture capital for a start-up — *"a real powerhouse"* Pierre-Gilles de Gennes would say of him. The director let them get on with it, but then, since the researchers were unable to obtain a conclusive result, he urged them to end the experiments.

Over the years, Pierre-Gilles managed to raise the level of research at the school. But he had higher ambitions. From the first day, his plan was to transform "PC" (Physics and Chemistry) into "PCB" (Physics Chemistry Biology), by introducing biology to the school. His early training at "NSE" (Normale Sciences Expérimentales) had left its mark, but above all he was convinced that the major discoveries of the 21st century would not be in physics or chemistry, but in biology. Once again, he met considerable resistance, but this time from outside the school. As its name suggests, the school is in fact owned by the City of Paris. Initially, relations were good, with the municipal director of education doing everything possible to accede to requests, but in 1977 the status of Paris changed. Instead of being run by the Prefect of Paris, municipal government passed into the hands of a mayor, elected by the members of the Paris Municipal Council. And the problems began soon after. The school cost money without bringing in votes. Although the amount was just a drop in the ocean of the municipal budget, the city came to see the school as a deadweight and started to query any expenditure on construction work or projects. *"The current financial situation requires all the City departments to look for savings"*, declared the director of education at the start of the school's management board meeting in 1985. Then, responding to the protests of research professors against the suppression of a lab technician's position, she declared in May 1986: *"The policy to reduce tax levels in Paris has led to a cut in the financial resources available to departments, and job reductions are one of the consequences."*

ESPCI Under Threat

In May 1987, against the background of global cuts in funding and jobs, Pierre-Gilles — dressed up to the nines, or at least wearing a tie for the occasion — went in person to City Hall in Paris, to explain the need to introduce biology into the school. He pleaded his cause, emphasising the support of renowned biologists like François Jacob, and concluded with a request for three additional posts (a professor and two lecturers) and the construction of new premises. The director of education laughed in his face: *"You're trying to squeeze us dry!"*, she exclaimed. Pierre-Gilles spoke not another word, while Lucien Monnerie, who was with him, vainly tried to argue. They encountered the same intransigent response from the director of the department. *"At that time, ESPCI had the same status as the city's nursery schools and had no independence. A form had to be filled in for every single order, then checked by the education department, the accounts department, etc. It was almost more complicated to buy a box of thumb tacks from the local shop than a microscope costing millions, because no one would accept an order form for a box of thumb tacks. To exaggerate slightly, you could buy a ton of pens, but not a single pen. It was impossible"*, explains Jean Léoni.[8] With more experience of the budgets of nursery schools than of research institutes, the director of education was astonished by the sums involved at the school and even suspicious. *"You're using City funding to do research behind our backs"*, she accused. Faced with such ignorance, Pierre-Gilles changed tack and tried to arrange a direct

---

[8] Today, the school an is run by an independent board, and receives a budget from the city, which it can manage with greater autonomy than in the past.

meeting with the mayor, Jacques Chirac, through the good offices of Alain Devaquet, professor of chemistry at Jussieu and mayor of Paris's 11th district. The latter promised him an appointment, but nothing happened. Time passed. Things went from bad to worse. The city was considering moving the school to the suburbs and selling the premises. There were suggestions of a change in its status (*"It could simply become a public-private company [with big industrial corporations]"*, suggested the director of education), or finally getting rid of it completely on the grounds that higher education was the responsibility of central government. *"Find yourself a buyer"*, was the blunt suggestion.

In despair, Pierre-Gilles and his directors tried to transfer the school into state control, making it part of the National Education system. *"In a letter of January 17, 1986, Mr Chevènement, Minister of National Education, in reply to a letter from Mr Chirac, Mayor of Paris, stated that he was in favour of (...) [the] conversion [of ESPCI] to a public establishment (...). The Ministry could finance the development of the school"*, crowed the Paris director of education. They drew up plans to merge with the Paris Higher National School of Chemistry, then with the École Normale Supérieure, holding multiple meetings with the Ministry of Higher Education, but it all came to nought. *"The malaise was profound"*, acknowledged Pierre-Gilles de Gennes. The school was sinking into the mire, from which only a Nobel Prize could save it, like a miracle that no one believed in any more.

And indeed, as soon as his Nobel prize was announced in 1991,[9] Pierre-Gilles resolved to use it to defend the cause of the school. He agreed to give a press conference and a live interview on TV news, provided that both events took place at the school. He continued to maintain this policy, using every possible public appearance to promote ESPCI. His strategy paid off. On November 27, 1991, the National Education Minister Lionel Jospin visited ESPCI, on the recommendation of his adviser Claude Allègre, an old acquaintance of Pierre-Gilles. It was a long time since the school had received so many honours. The theorist felt renewed hope. Nevertheless, he was

---

[9] See Chapter 13.

careful not to seem too favourable either to central government (left-wing), or to the Paris administration (right-wing). *"The political context was quite tricky. The school essentially depended on the municipal authority, but also on central government for a certain number of things. So we had to manoeuvre with a degree of flexibility. I always maintained neutrality in my dealings with both"*, he explained. His scientific director, Jacques Lewiner, managed to re-establish good relations with the city administration. The plans for a move were shelved. The school was saved. That was a first step. But it would take a second Nobel Prize to achieve more.

The Nobel Prize awarded the following year to Georges Charpak, a researcher at ESPCI, seemed like a miracle. The City of Paris suddenly discovered that it was sitting on a goldmine. It was actually a stroke of luck, because Georges Charpak had in fact spent his entire career at CERN and his secondment to the school, a purely administrative measure, was very recent. *"Charpak, who had retired from CERN, asked me to accommodate his small start-up at the school, where he was an associate professor. I simply wrote a letter, but it did the school a lot of good, since he won the Nobel Prize"*, chuckled Pierre-Gilles de Gennes. As a result of this second Nobel, the school could suddenly do no wrong in municipal eyes. *"Two Nobel prizes in two years! The politicians became very friendly. Before, we were up against a brick wall, but those two Nobel prizes opened a lot of doors. We managed to get funds to start construction work and introduce biology into the school. The years 1992 to 2000 were a happy time for the school"*, he recalled.

At the ceremony organised in honour of Charpak's Nobel Prize, Jacques Chirac finally set foot on the premises and Pierre-Gilles welcomed him with the remark: *"I might be slightly offended that you didn't come here for my Nobel Prize."* Then he showed him around the school's dingiest labs, just to show how needy the researchers were. Jacques Chirac spotted the manoeuvre, but took it in good part: *"I didn't get the chance to visit your smartest labs"*, he quipped.

In any case, major construction work began in 1994 on new labs, a 350 seat lecture hall, a dedicated space for popular science, etc. And Pierre-Gilles had something to celebrate: he finally had the green light to introduce biology into the school! The first professorial post was created in 1994. It had taken 20 years, but the physicist had finally succeeded in turning "PC" into "PCB" — something else he owed to his Nobel Prize. At a tennis tournament at Roland Garros, he would subsequently meet the director of education who had summarily turned down all his requests: she was all sweetness and light with France's favourite Nobel prizewinner. He responded politely, but not without bitterness.

At the school, the introduction of biology had a seismic impact, all the more powerful in that Pierre-Gilles recruited Jean Rossier, a neurology specialist, and then Serge Charpak (son of the Nobel prizewinner), a specialist in physiology (after the allocation of a second professorial chair in biology), rather than researchers in biophysics, biochemistry or molecular biology, as everyone expected. The research done by Jean Rossier was both fundamental, focusing on the organisation of the neocortex, and applied, since he was behind the development of a genetic analysis technique based on a single neuron. On the teaching side, Pierre-Gilles wanted students to understand the difference in approach between research in biology and in physics, and designed the curriculum accordingly.[10] *"Biological systems are complex and cannot be precisely resolved. The best the researcher can do is to test operational hypotheses"*, he explained. In order to fit biology classes into the timetable, he had to reduce some courses, which once again prompted resistance from certain professors. Pierre-Gilles simply ignored the opposition from what he called the *"traditionalists"*.

However, levels of discontent began to rise, as Pierre-Gilles gradually stopped holding meetings of the research and teaching committees. He had always been somewhat grudging about them, but now he chose to give up the hassle of meetings altogether and, since no one made a fuss, began to take more and more decisions solely with his directors, Jacques Lewiner and Jacques Duran, now director of studies. The professors felt excluded, for example only being informed of appointments after the event. For many of them, it

---

[10] The first year was spent on the description of cell function, the second year on the move from cell to organ, then the third year on the description of organ function.

was an unacceptable step back, a return to the regime of the mandarins. *"Even transparency had gone"*, sighs one of them.

Fortunately, his life as a director was not just an unending wrestling match with the school's research professors. *"There were good times. One day, a cleaner turned up holding a cannon shell that she had found in a closet — it must have been there since that time of Langevin, who was a navy consultant, and everyone had forgotten it. Panic! We had to call in the explosives experts. It was interesting to see who kept their cool, and who lost it. Later on, at the time of the terrorist attacks in Paris, someone found some bottles with a strange liquid in them. Once again, half the people cleared out. I gently unscrewed one of the bottles, sniffed and... it smelt of soap! They were bottles of detergent"*, chuckled Pierre-Gilles de Gennes.

In 1996, the movie director Claude Pinoteau was preparing a film adaptation of the successful play *Les palmes de Monsieur Schutz*, about the discovery of radium by Pierre and Marie Curie at ESPCI. His team contacted Jacques Lewiner to get access to the archives and do an accurate reconstitution of the lab where the Curies worked (the attention to detail in the film was so great that Marie Curie's writing was reproduced on a blackboard). *"Why don't you get de Gennes and Charpak to play the role of coachmen?"*, Lewiner suggested. So there they were, a few months later, at the Boulogne studio. *"To begin with, we spent half a day getting made up and changing into contemporary costumes. Then shooting began. We had to manoeuvre two enormous draughthorses and get them to stop at a precise spot: going forward was fine, but every time there was a retake, we had to get them to reverse. Things began to get tough! And there were six or seven retakes... We spent half a day shooting the scene, all for a couple of lines! Still, it's a good memory"*, recounted Pierre-Gilles de Gennes. The two Nobel Prize winners appear in the movie alongside Philippe Noiret, Isabelle Huppert and Charles Berling.

Taking Stock

Pierre-Gilles de Gennes had only intended to spend a few years at the head of the school. In the end he stayed 26 years. As soon as retirement started to loom, he began to look at possible replacements, immediately ruling out certain professors who might have fancied themselves for the job. His choice fell on Jacques Prost, to whom he described the job in idyllic terms: *"You'll see, one morning of administrative work a week and a few occasional meetings"*, he assured him. Initially not tempted, Prost eventually accepted. Of course, the position would prove much more demanding than expected. Pierre-Gilles could slide out of many obligations thanks to his reputation and the unfailing support of the directors and the general secretary, advantages on which his successor could not necessarily rely.

When Pierre-Gilles stood down as director of ESPCI in 2003, he could be proud of having saved the school, through his Nobel Prize, and of having improved it considerably, but it had taken him 25 years to do so. *"He was obsessed with elevating teaching and research at PC to the very top rank"*, commented a president of the alumni association. The physics research labs were now amongst the best in France. Young ESPCI graduates were still rated very highly. He had not accomplished all the reforms he dreamed of — some were unattainable: it was impossible, for example, to touch the entry exams, though he had very harsh views on the existing entrance system. *"Although director of the school, I am personally convinced that the current system of entry exams is deeply damaging. Every time a scientific competition is set up, teaching programmes become set in a mathematical mould and other qualities go by the board. (...) Unfortunately, the competition system is embedded in our culture, and it will take a long time to return to common sense"*, he wrote, adding that he found it *"scandalous that, on the grounds of having passed an exam at the age of 20, people have lifelong rights on society"*.[11] He did not undertake any structural reform at the school, despite a suggestion by Lucien Monnerie not long before his departure: *"The director remains in post until retirement. Shouldn't we introduce a term of office, for example a five year rolling contract?"* *"It's a good idea"*, he replied. *"I'll talk to Jacques Prost about it."* The final reform of his directorship was nipped in the bud. On one occasion, where a physicist was bold enough to accuse him of despotism, of betraying his youthful ideals, he candidly replied: *"The school's structure is archaic: I reap the benefits."*

---

[11] *Les objets fragiles*, Plon, 1994.

Chapter 13

A Different Kind of Nobel

On the morning of Wednesday, October 16, 1991, Pierre-Gilles de Gennes left ESPCI slightly earlier than usual and walked to the Collège de France. It was lunchtime, just before noon, and the lab was empty. The theorist was working in his office when the phone rang. He picked up. The caller introduced himself: *"I am Carl-Olof Jacobson, secretary of the Swedish Academy of Sciences. I am pleased and honoured to inform you that the Academy has awarded you the Nobel Prize for physics."* Pierre-Gilles was dumbstruck, then stammered his thanks. He hung up, left his office and ran into his colleague David Quéré. *"Ah! David, I've something to tell you: I've won the Nobel Prize"*, he said, before adding: *"What a pity that my mother's no longer with us."* Then, beaming, he rushed into the corridor, arms aloft, towards Madeleine Veyssié: *"I've got it! I've got it!"* he cheered. She found a bottle of champagne — warm, but no one cared — and celebrated the event with Liliane Léger and Anne-Marie Cazabat. Then all the phones in the building started to ring: the news had just been made official and journalists were already trying to contact the new laureate. Pierre-Gilles left the lab to have lunch with Françoise, as planned, in a small neighbourhood restaurant. But the press were already on the prowl in the Collège courtyard! He walked a little faster, eyes straight ahead. The journalists had no idea that the tall guy walking casually past them, holding a plastic carrier bag, was the Nobel prizewinner they were looking for. Pierre-Gilles managed to leave the courtyard without being spotted.

On his way to the restaurant, he mulled over the news and looked back over the years. Like many scientists, he had dreamed of the Nobel prize and knew that it was within reach. *"I had wondered if I had any chance of winning, and concluded that it wouldn't happen, because it is awarded for a specific discovery, and I didn't fit the profile. So I never got fixated on it, which was a good thing"*, he admitted. At the restaurant, he joined Françoise at the table. She was happy but also anxious: *"The Nobel Prize is going to put you in the spotlight. Our situation will be a problem"*, she feared. He was not going to let it bother him. Lunch over, they set off back to the Collège and, on arrival, found the courtyard swarming with journalists and TV trucks. This time, there was no escape: *"Monsieur de Gennes! How do you feel?"* He answered that the important thing in life is not awards, but... to see your children grow up! It was a reply, somewhat out of character, that would be printed in newspapers all over the world.

When he managed to get back into his office, he made two phone calls. The first was to Annie, at the *Boudin sauvage* in Orsay, who congratulated him joyfully. *"We gave everyone at the restaurant free champagne today to celebrate!"*, she said. The second call was to Jacques Lewiner, scientific director at ESPCI. Pierre-Gilles had already decided to use the leverage of his Nobel prize to extricate the school from its difficulties, and wanted his scientific director's agreement to schedule a late afternoon press conference at ESPCI.[1] It was only after he had hung up that was ready to speak to the journalists, who would spend the next few weeks filing through his office. His life as a Nobel winner had started.

At the school, an amazing explosion of joy greeted the radio announcement of the Nobel prize. Spontaneously, everyone spilled out into the courtyard: students, teachers and the whole staff. Within a couple of minutes, phones started ringing in the offices, as

---

[1] See chapter 12.

they had at the Collège. Five minutes later, the first journalist was there, leaning on poor Yvette Heffer, Pierre-Gilles' secretary, to give the name of the restaurant where the hero of the hour was having lunch. But Yvette Heffer stood her ground and said nothing. Jacques Lewiner became self-appointed press agent, and started filtering the journalists' requests, supported by researchers who stepped spontaneously into the breach. *"One sheet per request, with the name of the journalist and the paper, together with the nature of the request"*, he told them. The requests piled up. *"I expected a tidal wave, but what happened was beyond my wildest dreams"*, he reported. Technicians came and went in the courtyard, unwinding cables and connecting equipment. The dozens of journalists had to bide their time, waiting for the new star. When he finally arrived, in late afternoon, he created an amazing stir: applause rang out everywhere. The journalists and photographers jostled to see him. *"I remember an amazing party atmosphere"*, he recalled. The winner posed in the school's most photogenic location — a slightly fusty chemistry lab, previously picked out by Jacques Lewiner. He played the game and even agreed to wear a white coat — probably for the first time in a 40 year career. Then he gave his press conference in an old lecture hall with worn seats, crammed to the rafters. At 8 p.m., the tension rose a notch. In a few minutes, he would be interviewed live on prime-time evening news. He was a revelation: easy, smiling and relaxed, he seduced millions of television watchers.

The next day, despite demands from every side, he decided not to change his timetable and went, as planned, to a seminar with *Rhône-Poulenc* researchers near Lyon. However, he quickly realised it would not be just another day. On the platform at Gare de Lyon, journalists were lying in wait to join him on the high-speed train. Then, when he reached his destination, he was given a triumphal welcome. Finally, he discovered that the world's press was calling him *"the Isaac Newton of our time"*. He initially put the comparison down to a journalist *"with an acute attack of hyperbole"*, but on learning that it came in person from Carl-Olof Jacobson, secretary of the Swedish Academy of Sciences, he attributed it to *"Nordic lyricism expressed by Swedish academicians"* and rejected the comparison.

"Newton's stature was far greater than that of today's scientists. He invented the telescope when he was just 18. At 20, he understood interferential optics and, a few years later, gravitation and planetary motion. But he was a very difficult character, a dyed in the wool individualist", he argued.[2] Throughout the day, the tributes rolled in: *"He is an example to all French scientists"*, declared Hubert Curien, Minister of Research and Technology. *"The Nobel Academy has made an excellent choice"*, judged Louis Néel, himself a winner in 1970. *"He deserves the Nobel Prize at least five times over"*, enthused Pierre Bergé, department head at the CEA. Amidst this chorus of tributes, French President François Mitterrand and Mayor of Paris Jacques Chirac remained strangely silent: neither came to congratulate him, despite the fact that Pierre-Gilles was France's first Nobel Prize in physics for more than 20 years.

Telegenic Scientist

The television channels, aware that they had found themselves a star performer, kept a tight hold on him. *"We gradually discovered the world of television..."*, he commented, evasively. In fact, since he had agreed to take part in both *Dimanche, 20 h 10*, a programme fronted by Jean-Pierre Elkabbach on France's Channel Five and *L'heure de Vérité*, presented by François-Henri de Virieu on Channel Two, he had unwittingly triggered a major brouhaha, with both broadcasters battling for precedence. *"It's inconceivable that he should go on Elkabbach first!"*, groaned *L'heure de Vérité*, but Pierre-Gilles, having given his word, made his first TV appearance with Jean-Pierre Elkabbach. As he had decided to show a small physics experiment in front of the cameras, he turned up with a handful of test tubes. *"Not dangerous, I hope?"*, queried a cameraman. *"It's going to explode"*, chortled Jacques Lewiner. The programme went brilliantly. It was simple: Pierre-Gilles stole the show. *"Dazzling"*, according to *le Journal du Dimanche*. *"He radiates intelligence."*

---

[2] *Le Monde*, October 23, 1991.

From then on, every TV channel wanted him and the physicist went along with them, going from one programme to another, usually with an experiment in his pocket. For example, on Jean-Marie Cavada's Channel 2 programme *La Marche du siècle*, he shook a container full of foam in front of the cameras and the foam strangely disappeared. He explained that, despite its apparent simplicity, it was a phenomenon that was still not understood. He showed a pine cone on Anne Sinclair's show *7 sur 7* on TF1, paying tribute to two French physicists who had just explained its structure. He constantly turned the spotlight on other physicists — in particular his collaborators, emphasising that he would not have achieved anything without them — as if he wanted some of his glory to reflect on them. The only show he went to empty-handed was Bernard Pivot's *Apostrophes* on Antenne 2, but here he was surrounded by his guests, his friends Franco Fido (a literary researcher he had met at Berkeley in 1960) and Jean Jacques (director of chemistry at the Collège de France), Pierre Perret (adored by Annie de Gennes), and Anouk Aimée, idol of his youth. *"I have very good memories of the programme with Pivot. I like the man"*, he said. In any case, at every appearance he seemed delighted to be there and always at ease, holding or scratching his head, like the badly raised boy he was not. And like him, the viewers enjoyed it, amazed that the *"Newton of our time"* could speak so accessibly. His clarity, his modesty and his enthusiasm came across, to the point that the expression the *"de Gennes effect"* was coined to describe the fascination he exercised over *"the public as much as his colleagues and students"*.[3] The columnist Pierre Georges captured this fascination accurately and concisely: *"He is someone who you immediately want to be friends with, or whose disciple you want to be, just to enjoy the unusual privilege of becoming clever for a moment"*.[4]

If Pierre-Gilles — usually so stingy with his time as to avoid greeting his colleagues at ESPCI — was so generous with the media, it was not to warm his ego under the studio lights (in 1992, he declined a

---

[3] *Le Monde*, October 23, 1991.
[4] Billet de Pierre Georges, *Le Monde*, December 7, 1991.

proposal by Hervé Bourges, Chairman and CEO of the French TV channel Antenne 2-FR3, to direct a scientific broadcast), but firstly to help his beleaguered school, and secondly to improve the image of science. Through the Nobel laureate's TV appearances, the French discovered *"handmade physics"*, a science without equations that discussed day-to-day objects like sand, glue or water droplets, quite an achievement for a theorist at home with the most abstract formalism. He also used his media profile to put across a *"few messages"* on subjects such as education and the environment.

Nobel Spokesman

The Nobel Prize in fact bestowed powerful leverage. And all those who approached him in the hope of linking the name of Pierre-Gilles de Gennes with their cause knew what they were doing... The physicist overcame his initial reservations about *"petitionitis"*, as he called it, and committed himself on several occasions, for example signing the appeal for peace in Croatia launched by Linus Pauling (Nobel Prize for chemistry in 1954 and Nobel peace prize in 1962) in November 1991 — the first cause he espoused after the Prize — then to call for greater democracy in Cuba, the release of Bulgarian nurses, etc. Whilst interventions like this did not make waves, others — such as his ideas on education — triggered heated polemics.

"Since selection through mathematics facilitates the preparation and correction of exams, and also encourages cramming, both teachers and students like it. However, students become cut off from reality and run the risk of remaining so all their lives", he explained, pleading for a more pragmatic education that would teach students to express themselves, observe and do things with their hands. He proposed that from an early age pupils should get used to giving presentations in class, that children should be taken *"birdwatching"* and teenagers *"work in a garage"*, in order to experience the reality of the working world. When he heard it said that experimental equipment was expensive, he blew up: *"For the cost of a single computer, you could have a whole class doing experiments for a year!"*, at a time when the "IT for all plan"

launched by Laurent Fabius in 1985 (with the aim of putting a computer in every classroom) was nearing completion. Pierre-Gilles also advocated reviving *"attention to quality"* from nursery school onwards, citing the example of Japanese children who spent years learning the calligraphy of the kanjis, whereas *"on the grounds that they mustn't be traumatised"*, a French child would get as much praise for a mere scribble as his classmate for well executed work.

He had always denounced the excessive emphasis on theory in France's education system, based on *"the dictatorship of mathematics"*, but with the Nobel Prize he now had a receptive audience. His opinions, which also skewered teachers — or some teachers and some teaching practices — were not to everyone's taste. Teachers and school inspectors responded virulently, either in letters or in the columns of the newspapers. He continued to repeat his views tirelessly, aware that he was attacking *"ingrained habits"*, but in the hope that one day he would not only be heard but listened to.

He also triggered a polemic on the occasion of the Rio Earth Summit in 1992 by joining 70 other Nobel laureates in signing the Heidelberg appeal protesting against the emergence of *"an irrational ideology which is opposed to scientific and industrial progress, and impedes economic and social development"*. The signatories were immediately accused of having had the wool pulled over their eyes by industrial lobbies opposed to the Summit's environmental objectives. But Pierre-Gilles had never disguised his views on the environment. Without denying or minimising ecological deterioration, he decried what he called *"the religion of environmentalism"*, based on a cult of nature and the *"naive belief"* that everything that is natural is good. He issued numerous warnings about the nostalgia for the past sometimes expressed by ecologists, convinced for his part that the solutions to environmental problems could only come from technical progress, not from *"going backwards"*. So the Heidelberg appeal was entirely consistent with his personal views.

He also felt that this new religion of environmentalism was based on the exploitation of fear. This is why he came out against the destruction of experimental GM crops. *"This fear of the new has caused a famine in East Africa: the population refused to eat the maize that was sent to them. (...) You would think that we were still living in the Middle Ages, at the time of the great fears!"*, he raged,[5] pointing out that some of the plants destroyed produce proteins engineered to combat cystic fibrosis.

In general, he found that the approach to environmental problems was wrong. For example, he was irritated by the fashion for paper bags and glass bottles, which "sound green", without necessarily being so. *"The glass industry consumes much more energy than the plastics sector and, not so long ago, the paper industry was one of the biggest polluters"*, he explained. Out of the same desire to counter environmental myths, he signed a newspaper leader entitled *"Why biodegradability is not a panacea"*, writing:[6] *"Parliament, in its legitimate desire to protect the environment, is planning to ban plastic supermarket bags from 2010. However, bags will be permitted as long as they are biodegradable. It was even suggested that this policy should apply to all packaging: salvation through "biodegradability!"* He attacked both the message of biodegradability — *"Throw me away, I just disintegrate"* (when in fact, a biodegradable bag takes several months to decompose) — and a policy that entrenches *"the throwaway society"* to the detriment of *"re-use"*. In addition, the manufacture of biodegradable bags uses large quantities of water and fertiliser. These were just the sort of scientifically unfounded *"measurettes"* that he objected to. They make people feel righteous, but ignore the real environmental problems, which were, in his view: overpopulation, water shortage and energy production.

On this latter point, his position was clear: *"We need to live with nuclear energy, there is no way round it."* He was convinced that nuclear power should even be expanded to combat climate change, and signed a newspaper article appealing to *"our fellow citizens and leaders to pursue a deliberate and resolute policy both to save energy and to develop nuclear and renewable energy"*,[7] as well as backing the manifesto of the

---

[5] *Capital*, December 2003.

[6] He signed this leader in *Le Figaro* on December 22, 2005, with Guy Ourisson, a chemist who was President of the Academy of Sciences and Richard-Emmanuel Eastes, a scientist at the École Normale Supérieure.

[7] Leader by Georges Charpak, Pierre-Gilles de Gennes and Jean-Marie Lehn and others, in *Le Figaro*, December 26, 2006.

Sauvons le climat [Save the Climate] association that promulgated this approach.

Before taking a position on any question, he would always take the time to study the pros and cons enough to form his own opinion. Otherwise, he would refrain from stating a view — for example, on cloning, he would always refer to the conclusions of the National Consultative Committee on Ethics — or else confine himself to what he knew. For instance, when the scientific community protested at the title of doctor being conferred on Élisabeth Teissier in 2001, describing her PhD thesis presentation as a *"pseudoscientific farce"*, he did not directly attack the thesis (being unfamiliar with its content) in the letter he sent to Jack Lang: *"Dear Minister, Élisabeth Teissier has just submitted a sociological thesis on astrology at the University of Paris 5. I believe that the sociological aspect of this belief (and of this trade) would indeed merit objective study. However, I am concerned by Madame Teissier's suggestion regarding a future course on astrology at the University. In our era, infested with sects and pseudo-sciences, I would find the introduction of such a course deeply shocking."*

When asked to express an opinion on subjects other than his work or research and education policy, he would stress that his opinions were entirely personal, and that he made no claim to oracular status. Nonetheless, his words could sometimes sound prophetic, as when he wrote: *"All in all, the society that we see emerging is a society of comfort. Not dominated by technicians, as Jules Verne imagined, but impoverished in its culture, as he thought. Enjoying advantages over the Third World that are totally unjustified"*,[8] or when he declared: *"In some thirty years, the world will be dominated by Southeast Asia"*.[9] Later, on the Iraq war, he condemned the US occupation saying that it would *"eventually create a breeding ground for pan-Arabic insurrection, far greater than anything we know. (...) We, as Westerners, are going to find ourselves at the centre of an immense, incoherent and lasting conflict."*[10]

---

[8] *Les Échos*, February 1, 1995.
[9] *Le Télégramme de Brest*, December 13, 1991.
[10] *Le Monde*, February 14, 2003.

However prudent he tried to be, some accused him — on occasion vehemently — of giving his opinion on anything and everything. *"To see the PGG of our youth — who preached mobility and inspired trust because he matched his words with action — turn into the intransigent boss of ESPCI, is sad enough. But to hear you yesterday, on Heure de vérité, asking workers to give up the right to strike out of solidarity with the unemployed, when for the last sixteen years you have held two posts, from both the Government and the City, is just too much!"*, wrote a former collaborator. *"To see yet another theorist give his opinion on everything without knowing the truth is a spectacle to which I have become accustomed. But in you, this level of arrogance saddens me"*, he replied.

Professor Eugène

These polemics were undoubtedly the inevitable price of fame. For Pierre-Gilles had become a celebrity, recognised and hailed in the street. *"Wherever he went, people stared!"*, recalls Annie de Gennes. He felt *"quite touched"* by the marks of friendship, even finding certain benefits in fame. *"One evening, I rushed into a pharmacy, because I had lost my asthma medication, which is only available on prescription. I didn't expect much. But fortunately, they recognised me and I was able to get it"*, he chuckled. If he was only rarely bothered, it is also because he made efforts not to be. *"People would sometimes stop him in the street to ask for his autograph. He always gave it. But he often made sure that he wasn't stopped... by walking very fast!"*, laughs Phil Pincus.

The new Nobel laureate now received mail by the sackful. He was so famous that a letter sent from Africa to *"Professor Eugène — Great scientist — France"* actually reached him. He was keen to respond to all these letters, so would get up two hours earlier to be in his office by 6 a.m. He would go through each letter and note his answer in the margin, in pencil, with an eraser nearby, just as he did with reports and articles. His reply usually consisted of a few lines, even a few words — *"impossible, sorry"*; *"yes, make appointment"* — that Yvette Heffer, his faithful secretary, would then type up. He would sign, and hey presto, it was all done the same day. Most of the correspondence consisted of congratulations — *"The prize prompted many childhood friends, some I hadn't seen for 40 years, to get in touch. It was moving."* For a laugh, he would pin the letters sent by his most ardent detractors to his office door. He also received a large number of scientific questions, which he tried to respond to, though he didn't always have answers. When a retired doctor wrote asking about the adhesive power of holothuroidea filaments, he sent a note apologising for his ignorance of sea cucumbers. A wine grower in Volnay asked him whether a litre of water heated in a microwave cools at the same speed as a litre of water heated on the hob. Intrigued, the physicist ran the experiment and observed that the microwaved water cools faster. He and his physicist friend Tom Witten came up with several hypotheses to explain the phenomenon. A young physicist also asked him why the footprints of astronauts leave such clear markings on the moon. But he also received declarations from *"lonely women"*, messages from harmless nutcases claiming to have solved the greatest mysteries of science or from oddballs outlining their latest inventions. He sent them all a kindly reply, explaining to the eccentrics that the validity of their invention *"remains to be proven"*. Some letters led to meetings — for example, he made an appointment with a female admirer whose letter had "*moved"* him — and to correspondences that lasted for years.

This tidal wave of mail also included dozens of invitations every week to attend inaugurations, debates, lectures, etc. His presence became highly prized, sometimes advantageously rewarded. *"With me as impresario, him as the star, we could have made a fortune"*, jokes Jacques Lewiner. With all this demand, he had to turn down nine in every ten proposals, beginning with all the society invitations. But he couldn't avoid them all. So he would often manage to slip discreetly away, despite being the centre of attention! He would sometimes find himself involved in initiatives that seemed far from his area of expertise, for example on the honorary committee of the *Grands ateliers de France, "a selection of traditional firms of the highest craftsmanship in the field of art, decorative art and the art of living"*; at the opening of a conference on meteorology at the Royal Saltworks at Arc et Senans or at the Congress of the *French Dentofacial Orthopaedic Society* in Saint-Malo... He was also very much in demand to write prefaces — something he was happy to do if the book interested him, unless the publisher made him revise his text. *"I am not in the habit of being so firmly guided in my writings. But I have made two corrections. If your publisher is not satisfied with the result, let us drop the whole idea"*, he wrote angrily in 1996 about one of his prefaces. He also put his name to the prefaces of books as varied as *Preventing corrosion and scaling in water distribution systems; France's most beautiful treks; God equal to man, a critical rereading of christianity*, not to mention innumerable scientific tomes. Similarly, he was happy to lend his name to initiatives such as the Sand Museum near Sables d'Olonne, the Scepticism Laboratory at Nice University (study of paranormal phenomena) or the science journalism section of the Higher School of Journalism in Lille. He also on occasion sponsored humanitarian bodies such as *Frères des hommes* or *Secours populaire français*: *"We have unemployment of 10%. And even more families worried about paying their bills. We need to act. Obviously, at the level of Government and the economy, but that is not enough. There are people who have fallen through the net, and amongst them a good number who want to make good, but lack self-belief. These are the ones for whom the Secours Populaire exists, giving them a helping hand at difficult times and giving them hope. This concerns everyone in France. By helping the Secours Populaire, they give our country new hope"*, he wrote in 2004.

French School Tour

A few weeks after the Nobel announcement, he received a letter that caught his eye. A group of schoolchildren who wanted to meet him had pooled their resources to buy him a train ticket. He was touched by the gesture. For a number of years, he had wanted to talk to young people about science, instead of just bemoaning the loss of interest in the sciences. However, as the director of a higher education establishment, he could not speak to schoolchildren without being suspected of recruiting for ESPCI. This letter persuaded him to pick up his pilgrim's staff to argue the cause of scientific research with young people. He counted on the prestige of the Nobel Prize to ensure a hearing.

The physicist did not expect his talks to be such a success: every lycée in France and Navarre wanted to hear him. The venues were sometimes so full that he had to step over the legs of youngsters sitting on the floor to reach the podium. He made the effort to deliver one lecture a week for almost two years, all over the country. *"We had to find trains that arrived just before the lecture and left immediately after"*, recalls Yvette Heffer. That way, he could largely dodge the inevitable receptions organised in his honour. *"All the same, they took almost a whole day by the time I got there and delivered the lecture"*, he recalled. Which was no trivial matter, given his timetable.

In setting out on this tour of the schools, Pierre-Gilles mainly wanted to start a dialogue with schoolchildren and create the best possible conditions for the youngsters to feel at home and express themselves. So to begin with he asked teachers not to attend his talks, but their insistence quickly proved too strong. He preferred big venues where the students would feel safety in numbers and would find it easier to ask questions. So he was perfectly happy to speak in stadiums, like the one in in Orthez, filled with busloads of students shipped in from the surrounding towns. A local politician wanted to preface the lecture with a few words. The physicist disapproved, but said nothing. When the politician saw the thousands of overexcited youngsters in the stadium, he went white. *"Please, go ahead"*, he urged, missing the Nobel prizewinner's ironic grin. He stepped forward and effortlessly captured the attention of his young audience. On another occasion, in a crammed marquee at Saint-Denis, he instantly defused the tension that arose when the projector broke down. The teachers were afraid that the kids would start heckling, but Pierre-Gilles sat down with them and started a light-hearted discussion while the apparatus was repaired, before picking up his lecture where he had left off.

His tour also took in the Saint-Louis lycée, where he had not set foot since 1952 when, as a student at the École Normale Supérieure, he gave test papers to future candidates. He found the school less dark than he remembered. He also gave a talk at Lycée Saint-Cloud, where his son Mathieu was a pupil. The boy was somewhat dreading the prospect, probably afraid that his father would not make the grade. As usual, however, the lecture went very well. *"I have never faced aggressive questions. There are really questions about everything, about me, about how much I earn, etc., slightly silly questions as well — is Anne Sinclair as beautiful in real life as she is on TV? But I have never had to deal with hostile questions, except perhaps once, in Antibes: a youngster who wanted to show off, but it didn't go far"*, he recalled. His lectures were not restricted to mainland France, since he also spoke in Martinique in March 1993, where he had taken his four young children for the spring vacation — *"We did a tour of the island and I met Aimé Césaire at the town hall. A really great experience"* — and also in England and India.

The literary editor of a publishing house suggested that he should turn the lectures into a book. Pierre-Gilles was in favour of the idea but did not want to do it himself and turned the writing over to Jacques Badoz, his former scientific director at ESPCI. *Les objets fragiles* came out in 1994, a collection of lessons on soft matter and thoughts inspired by his conversations with schoolchildren. The book was a great success, selling 70,000 copies in France, as well as versions in several languages. When Jacques Badoz was asked about his contribution, he mischievously answered that he had written the captions to the illustrations (the illustrations do not have captions). Pierre-Gilles considered writing another book, a whistlestop tour of research in France, inspired by the French classic *Tour de la France par deux enfants* (1877) [Two children's tour of France], but the plan never came to fruition.

Behind the Scenes at the Nobel

There was a two-month gap between the announcement of the winners and the Nobel awards ceremony. The names of the recipients are a well guarded secret until the official announcement. Nonetheless, it is likely that Pierre-Gilles had some inkling. A few of his colleagues had been approached by the Swedish Academy of Sciences (on behalf of the Nobel Committee) to prepare a secret report on his work (A13-1). Tongues had wagged and a rumour got around that the soft matter physicist was in the frame for the Nobel

Prize. That was not all. At the beginning of October 1991, Pierre-Gilles happened to be in Sweden giving a series of lectures. A few clues might have made him smell a rat. For example, he was invited to dinner by a chemist, a member of the Swedish Academy of Sciences, who got him to talk at length about his work, concluding: *"Ultimately, it's true that you're a physicist, not a chemist."* At the time, the remark passed him by, but it made sense when he learned that the Swedish Academy of Sciences had discussed whether he was not more a chemist than a physicist. Three days after this meal, he was unexpectedly — and without explanation — summoned, via the French Embassy in Stockholm, to a photo session. Something of a giveaway! He had entertained the possibility since 1988 that he might win the Nobel Prize, but in 1991, the rumour seemed to be becoming reality. In fact, on the morning of the prize announcement, on October 16, he had left the school earlier than usual, being strangely careful to tell his secretary that he was on his way to the Collège, instead of his usual vague *"see you later"*. Though he could not be certain that he would win, Pierre-Gilles at least suspected that he would be a serious contender.

That did nothing to diminish the emotion he felt at the news. *"He was very surprised when I called him"*, confirms Carl-Olof Jacobson. *"He had been in Sweden the previous week and suspected nothing, despite the photos we took for the press releases."* His pleasure was all the greater in that there were several distinctive aspects to the award. Firstly, it came on the 90th anniversary of the creation of the Nobel Prize, an occasion marked by particularly sumptuous celebrations. Secondly, he was the sole recipient, whereas the prize is often shared by two or three physicists. And finally — and most notably — he was awarded the medal not for a specific discovery, as is usually the case, but *"for discovering that methods developed for studying order phenomena in simple systems can be generalized to more complex forms of matter, in particular to liquid crystals and polymers."* In short, the theorist was rewarded for having demonstrated the existence of numerous analogies between solid-state physics and soft matter. One amusing detail: when invited in September 1990 by the Swedish selection committee to submit proposals for the 1991 Nobel Prize for physics, he had suggested two names, stressing that one of them *"had been (like Landau in the past) ignored up to now by the Nobel Committee, because of the sheer abundance of his theoretical discoveries"*: he was unwittingly preaching his own cause.

Great as was his pleasure at the award, the prospect of the ceremony itself was daunting. He quickly decided not to take his wife Annie or Françoise, but a few days after the announcement of the prize, asked Madeleine Veyssié, director of the laboratory at the Collège, to accompany him to Stockholm, along with Étienne Guyon, his former student at Orsay and now director of the École Normale Supérieure, Jean Jacques, director of chemistry at the Collège de France, the American physicist Karol Mysels, and his two great friends, the theorists Shlomo Alexander and Phil Pincus. He sent his guest list to the organising committee in Stockholm, provided his measurements for the borrowed suit and the matter was settled.

"All is but soap bubbles"

That year, some 130 Nobel laureates travelled to Stockholm to celebrate the 90th anniversary of the prize, amongst them the Dalai Lama, Elie Wiesel and Lech Walesa. However, it was the absence of the winner of the Nobel Peace Prize, Aung San Suu Kyi, under house arrest in Burma, that marked the ceremonies. Pierre-Gilles arrived in Sweden five days before the prize award. He had a busy schedule. He was met by a chauffeur and a bilingual escort, Eva Paulsson, assigned to help him throughout his stay. *"This very knowledgeable lady went with me everywhere and arranged everything. It was a detail, but it made life a lot easier."* Half an hour after landing, he was already at a reception, a foretaste of three days of press conferences, interviews, seminars and official dinners. Unable to avoid these obligations, he made the best of things, resigned himself to wearing a tie (borrowed) and enjoyed the opportunity to meet the writers Claude Simon and Gabriel Garcia Marquez. However, in any free moment he would slip away to admire the city's famous statues, guided by his little Swedish escort, who introduced him, for example, to the works of Carl Milles, a pupil of Rodin.

On Monday, December 9, the day before the award ceremony, he gave a lecture to a prestigious gathering of Nobel Prize winners, including the Frenchmen Jean-Marie Lehn and François Jacob, and the physicists Philip Anderson, Leon Cooper, Ivar Giaever, Alex Müller, Linus Pauling, Ilya Prigogine, John Schrieffer, Kenneth Wilson, etc., most of whom he knew well. He ended his lecture with a flourish by reciting a piece of doggerel: *"Let's have fun on land and sea/Fame brings naught but troubles/Riches, honours, this world's false fee/All are but soap bubbles"* — very apposite for the occasion...

The day of the ceremony arrived, Tuesday, December 10, the anniversary of Alfred Nobel's death, and began with rehearsals that lasted the whole morning, since the protocol had to be note-perfect. Pierre-Gilles then returned to the sumptuous harbourside Grand Hotel to get ready. Realising that he had no cufflinks, he made do with a couple of paperclips. He joined his friends in tails and evening gowns in the hotel corridor. They joked and had photographs taken before moving on to the gigantic Stockholm Globe Arena where the 4 p.m. ceremony was to take place. *"It was impressive, beautiful and majestic. More than 5000 people were there. The beams were decorated with thousands of flowers"*, recalls Madeleine Veyssié. The royal family was enthroned on a huge stage, along with the choir and the philharmonic orchestra, and all the former Nobel Prize winners present. The ceremony began. The new laureates filed solemnly across the immense hall towards the stage. As he went by, Pierre-Gilles winked at Madeleine Veyssié, Étienne Guyon and Phil Pincus, who were seated together. Once on stage, the laureates sat down opposite the royal family. When his name was called, Pierre-Gilles stood up and took a few steps until he stood before King Carl Gustav, who handed him the Nobel Prize medal in its case. After exchanging a few words, the physicist walked backwards to his seat, since protocol prohibits turning one's back on the King.

The ceremony was followed by a splendid dinner in the great hall of the City Hall. The royal family, accompanied by the new laureates, descended a majestic staircase, watched by the 1300 guests at their tables. *"I was looking around — the hall was superb — and unfortunately stepped on the dress of the princess in front of me. She gave me a furious glare"*, he recalled. He promised himself to be more careful and made his way without further incident to the table of honour, where he sat between Princess Christina, the king's sister, and Ingegerd Troedsson, President of the Swedish parliament. The dinner began. Dancers from the Stockholm ballet corps performed a few graceful steps, followed by dozens of waiters bearing the dishes. The menu consisted of nettle soup, marinated tartare of salmon with red paprika cream, roast duck breast with a root vegetable compote, and a dessert of vanilla ice cream.

During the meal, each new Nobel laureate had to give a short speech from an elevated dais. Pierre-Gilles had carefully prepared his own, inspired by the *Wonderful Adventures of Nils*, *"a story I loved when I was six. I can still remember the name of the goose leader."* He was half listening to the other winners' spiels, when his blood suddenly ran cold. One of them had just referred to the same story. His whole speech was ruined. *"I felt terrible. I absolutely had to find another idea quickly, whilst still making conversation with my table companion, Princess Christina."* It would soon be his turn. He tried to keep his cool. Fortunately, just as he was due to speak, he remembered *Cinderella*. In the time it took to reach the chair, he had fine-tuned another speech: how surrounded by so many princesses, he had the impression of being in a fairytale and hoped he would not be turned into a pumpkin when the clock struck twelve. The ordeal was over. He could sit down and relax. The reception finished with a ball, but he did not stay late.

On the following day, all the Nobel laureates were invited to have their say on the future of the world in a long television debate, which Pierre-Gilles — unsurprisingly — found tedious and pointless. However, the magic of the previous day continued a little longer, since he and the other winners were borne by horse-drawn carriage to the Royal Palace for a final dinner given by their Majesties, the King and Queen of Sweden. Lulled by the sound of horses' hooves on the cobbles, he marvelled to see the entrance to the Palace illuminated by a multitude of torches carried by children. The next day, he returned to Paris, after a tour of the Nobel Foundation and a final lecture in Uppsala.

The fairytale was over, but there were a few honours still to come. A big party was held at ESPCI to celebrate his prize, a colloquy at the CNRS and in February 1992, a ceremony in the great amphitheatre of the Sorbonne, attended by Lionel Jospin, Minister of Education. *"That day, I lent him my driver. In the past, he had always refused. How many times did I tell him: "It's practical for getting from one appointment to another in Paris: you can work in the car and there's no problem with parking." But he was never interested"*, records Étienne Guyon, then director of the École Normale Supérieure. Pierre-Gilles preferred to walk or use his own car, even occasionally the *Boudin Sauvage* van. In February 1992, he inaugurated the first street to bear his name, in Labège. Other Pierre-Gilles de Gennes streets would appear all over France, and a square in Barcelonnette and in Orsay, a stone's throw from his house. But the party that moved him the most was the one laid on by his wife Annie, who secretly invited all his old friends from the École Normale Supérieure to Orsay. *"That was a wonderful surprise!"*, he smiled.

Red Carpet

The Nobel Prize was an open sesame to some enviable privileges. For example, it unlocked the doors of the Villa Medici in Rome, where Pierre-Gilles and Annie stayed in March 1993, as guests of the French embassy. He was also invited to join the opera maestro Placido Domingo for dinner, after a concert given by the tenor on Mayor Principal Square in Madrid. The prize lent a new shine to his letters, which inevitably made his correspondents sit up and take notice. When he wrote to Arthur C. Clark, *"I have always been a great admirer of your books (...). This is just a note of thanks for having brought us so many dreams and so much laughter... But if ever you come to Paris, or I to Sri Lanka, it would be a pleasure to meet you!"*, signing *"Pierre-Gilles de Gennes, Nobel Prize for physics 1991"*. The author replied the next day: *"It's not every day you get a fax from a Nobel prizewinner!"*

Wherever Pierre-Gilles went from now on, the red carpet was rolled out. In 1996, he was welcomed with great ceremony in India, to give a lecture to the Alliance française, before riding around

Bandipur Park on an elephant and visiting the formidable Golkonda Fort. When he travelled with his family, he was offered *"comfortable accommodation"*. Invited in February 1995 to Mérida, capital of Yucatàn in Mexico, for example, he and his four young children were put up in a huge and magnificent architect-designed house not far from the Chicxulub crater (created by the 10 km meteorite thought to have wiped out the dinosaurs). On one side was a view of the sea, on the other wetlands carpeted with pink flamingos. They would go down to the beach by a long, steep staircase, Pierre-Gilles and his four-year-old son Marc amusing themselves by counting the steps. Similarly, invited to give a lecture in Istanbul in July 1995, he lodged in a glorious old Turkish house near Bodrum, with a fountain in the inner courtyard. The setting was equally gorgeous on the island of Tobago, in the South Caribbean, where he stayed in 1996, a stone's throw from the Caribbean Academy of Sciences. He went swimming with his children on the coral reefs or in isolated creeks. Then they moved on to Trinidad, where Pierre-Gilles was excited to visit the places described in V.S. Naipaul's *Miguel Street*, one of his favourite books, which describes the colourful life in one of the town's impoverished neighbourhoods.

He had not always travelled in such comfort. On his trip to China in May 1988, everything had gone wrong. *"We arrived at Beijing airport, but no one was there to meet us. We eventually found a hotel. We tried unsuccessfully to reach the organisers through the Embassy. I remember arriving at Tiananmen Square: we were hungry, there was a stall selling Peking duck, which we ate standing up. A few days into the visit, I was scheduled to meet a leading polymer scientist in the west of the country, but there was only one seat on the plane. "Your wife will come on after", I was told. I left and, once again, no one at the airport. I had no idea where I was. Finally, a Chinese guy arrived and checked me into a horrible concrete hotel. I was worried about Annie, who arrived after... an 18 hour train journey. The national anthem was played — it was the rule — as she got off the train, exhausted: we were very happy to see each other"*, he recalled. To recover from the trip, they allowed themselves two days in Hong Kong, lounging around the hotel's rooftop swimming pool.

After the Nobel Prize, such misadventures became a thing of the past. Wherever he went, he was greeted like royalty — in May 1994, a party was even organised in a village in a remote corner of Tunisia. He was met at the airport, housed in the best establishments. In 1997, he stayed in a luxury hotel near New York, where, in the morning, he was to meet his host, Chen Ning Yang, winner of the Nobel Prize for physics in 1957. When he arrived, the latter exclaimed: *"Wow, I'm going to enjoy having breakfast here. The view from this terrace is fantastic!"* Pierre-Gilles did not dare confess that he had already taken copious advantage of the hotel's gargantuan and delicious breakfast.

In April 1995, he was invited, with Annie, to an international conference on molecular cooking, a gathering of top chefs and scientists in the beautiful little mountaintop town of Erice, in Sicily. On the menu: lectures on mayonnaises, sauces and jellies, but also sumptuous meals. *"This time, Annie was the star"*, he claimed, but it was he who was given the honour of opening the colloquy, with a lecture that he entitled *"Thoughts of a food-loving scientist"* (followed by Raymond Blanc's *"Thoughts of a science-loving chef"*).

All these honours and privileges did not go to his head. He still spent his summers in the Alps, hiking with the family in all simplicity. *"I also went hiking with the photographer Charly Bayle, who showed me some wonderful sights. For example, we got close to an eagle's with chicks in it."* He continued to stay regularly with his friends Louis Blanchard and Loup Verlet, and also with Jean Léoni, general secretary of ESPCI, in Corsica, and in 1992 with Jacqueline Petit — the girl-friend of his youth from the internship in Banyuls — in Porto Vecchio, etc.

"Hello, Jacques Chirac here"

After his Nobel prize, he was courted even more insistently than before by industrialists seeking to entice him onto their scientific committees or their boards of directors. In 1976, the firm *Air Liquide* had offered him a place on their board, but had not finalised the agreement on the grounds that his civil servant status as a professor at the Collège de France was a problem. The difficulties miraculously vanished after the award of the Nobel, prompting the journal *Option Finance to comment*: *"Certain companies have distinguished themselves by appointing independent directors, such as the Nobel laureate Pierre-Gilles de Gennes at Air Liquide (…). Until they prove their value, these initiatives might be suspected of being no more than marketing ploys."* Pierre-Gilles also agreed to join the scientific committee of *EDF*, but turned down the offer from the pharmaceutical firm *Sanofi*, not wishing to accumulate too many posts. The CEO persisted, but made no impression. *"One Saturday morning, I was working in my office at ESPCI when I heard the phone ring. As there was no one to answer, I picked up"*, he recounted. *"Hello?"*, he said, but there was no answer. *"Hello!"*, he repeated impatiently. *"And then I heard: "Hello, Jacques Chirac here!""* — this was during his Presidency — *expressing his wish for me to join the board"*. Under this "considerable pressure", the physicist gave way and became a director of *Sanofi* in May 1999, rubbing shoulders with men like Thierry Desmarest, CEO of *TotalFinaElf* and Lindsay Owen-Jones, CEO of *L'Oréal*. *"I don't regret it, because I got the chance to meet Gérard Le Fur* (Author's note: a distinguished scientist and director of research at the time), *an amazing guy, incredibly knowledgeable"*, he commented.

His new status also raised his profile with the politicians. In 1993, while staying in Sendai in Japan, he received an urgent message asking him asking him to call the office of the French Prime Minister, Édouard Balladur. He was intrigued. He had only met the politician twice (first as part of a scientific delegation pleading the cause of the chemicals industry and then at a lunch organised with Vivendi CEO Jean-Marie Messier). So he had no idea what Édouard Balladur wanted, but nevertheless returned the call. The new prime minister explained that he was in the process of forming his government and had thought of him as minister of research. Pierre-Gilles was surprised, but replied without hesitation, after the appropriate expressions of gratitude, that he did not feel he had the right qualities. The Prime Minister persisted: *"What if we put you in charge of education as a whole? Would you be interested?"* *"Not in the slightest"*, he repeated.

Right-Wing, Left-Wing?

Pierre-Gilles de Gennes would never be a minister. Despite the many approaches — Édouard Balladur was not the only one to try to bring him on board — he refused political office and would only give his backing to two candidates, Jean Tibéri in the 2001 municipal elections and Jean-Pierre Chevènement in the 2002 presidential election. Jean Tibéri had done so much for ESPCI that Pierre-Gilles felt in his debt, and agreed to support his campaign, *"out of loyalty to the man rather than political conviction"*. He was also on the Honorary Committee for the "Paris 2000" campaign that Jean Tibéri set up to organise the millennium celebrations in Paris. By contrast, he would be more open in his support for Jean-Pierre Chevènement, going so far as to declare at a meeting that he shared some of the politician's ideas and that Chevènement had been a *"good research minister"*. Nevertheless, it was during the latter's term of office that he had said to a colleague: *"The left is going to destroy research."* His colleague replied: *"No more than the previous governments..."* *"Yes, more! They provide money, but don't care how it's spent"*, he retorted. *"In the 1960s and early 1970s at Orsay, Pierre-Gilles de Gennes claimed to be on the left, but his behaviour was not particularly left-wing... As a result, he was unpopular with certain scientists, but it was a passing phase, arising from the political dissensions of the time"*, recalls a fellow physicist. In the end, Pierre-Gilles supported one candidate of the right and one of the left, which shows how little he cared about the difference between the two. He was more interested in how effective the individuals might be.

He was more ready to commit to broad democratic causes, for example participating alongside Yannick Noah in "New Citizens' Week", a movement to encourage young people to register to vote in 1991. In the second round of the elections between Jacques Chirac and Jean-Marie Le Pen, he appealed to scientists to vote for the outgoing president: *"We, as scientists, call upon you to make May 5 a referendum in defence of the Republic and democracy. All that will happen if you abstain or spoil your vote is that you will increase the representation of the extreme right, so we call upon you to vote unhesitatingly for Chirac."* He also signed the *"one politician, one job"* petition in 1997, calling for the end of multiple political functions, on the grounds that *"the Republic is in crisis, its representatives becoming discredited, its values under threat"*.

Broadly speaking, apart from his commitment to democracy, he kept his distance from politics. Nevertheless, he came into contact with many of France's big political names, men like Alain Juppé, François Mitterrand, François Bayrou, Philippe Seguin, etc., but these encounters always left him with the enduring impression of speaking in a vacuum. *"I remember talking about mad cow disease to some very senior figures, at a time when the problem was unknown in France. I had heard it said, probably at Rhône Poulenc, that there were concerns about the use of cow cartilage and bones to make gelatine. I heard about it quite early on. But nobody listened to me, or it just went in one ear and out the other. It caused the authorities a lot of embarrassment when it came out later"*, he remarked. Indeed, when the BSE crisis blew up in 1996, the government was caught short and Pierre-Gilles could only observe that his warnings had been ignored. However, above all it was the failure of ministers to act on research and education that made him cynical about politics. He had sounded enough alarms, made enough approaches to ministers and their office directors — nothing happened.

The picture he drew of the situation was sombre: in his opinion, the problem with public research was not lack of resources, but *"the lamentable performance of the universities"*.[11] In fact, he felt that university laboratories behaved as if they were self-subsidising. *"The universities are run by a council on which the biggest labs are the best represented"*, he explained. *"Their objective is self-perpetuation rather than scientific renewal, so, for example, they recruit their own students, students who did their PhDs through the laboratory, rather than outside candidates. The result is that you get the same research topics over and over again, however obsolete. This system is killing university research."* He tirelessly reiterated his preferred solution. He argued for the universities to become more autonomous — which is also the aim of the reforms

----

[11] *Commentaire*, 108, 2004.

introduced in 2009 — but still to be subject to controls. What controls? He suggested, for example, that funding should be based on results. *"We should make every university department independent, then allocate funding not on the basis of the number of students entering the university, because this system leads to the familiar abuses — insufficiently stringent recruitment criteria, too easy exams so as not to lose students and to justify staff levels — but on the basis of the number of students who get jobs after their degrees. This would oblige department heads to go looking for ideas and young scientists to put their labs in the forefront of innovation"*, he suggested, pointing out that American researchers, who are required to compete for funding, are extremely imaginative in proposing new research subjects that are both fruitful and attract support.

Nobel prize or not, his recommendations fell on deaf ears. Just once, he had the sense that he might have got through: it was with Valéry Giscard d'Estaing, who received him at the Élysée Palace during his presidency. At the end of the meeting, the President actually asked him to find *"someone able to carry through through these reforms"*. Pierre-Gilles left him in an optimistic frame of mind, but was unfortunately unable to identify the right person. And that was the end of the matter.

Claude Allègre

Claude Allègre might not have been that right person, but, when he was appointed Minister of Education, Research and Technology in Lionel Jospin's government in 1997, Pierre-Gilles felt optimistic: *"Here is someone who will push through the necessary reforms"*, he thought. The two men were old friends. In March 1978, Claude Allègre, a researcher in geophysics, had invited him to give a lecture on instabilities in fluids at the Globe Institute of Physics, explaining: *"We like to invite nonspecialist speakers to talk on various subjects that have the potential to generate new approaches to geophysics or geochemistry."* It was an attitude after the physicist's own heart. Since then, the two men had found much common ground and frequently spoke on the phone.

From 1988, when he was an adviser to Lionel Jospin (then Minister of Education, Youth and Sport, subsequently Minister of

Education), Claude Allègre had introduced measures to stimulate research in the universities. These included a system for evaluating departments, overseen by Vincent Courtillot (a physicist at the Globe Institute of Physics and at the time director of research and doctoral studies at the Ministry of Education), which the soft matter specialist had warmly applauded: here was someone who was prepared to cut through the business-as-usual political niceties. *"I'm not here to see the hardware"*, he would say when arriving to inspect a lab. After a few years, however, Pierre-Gilles observed that the evaluations — honest and stringent as they may initially have been — became more consensual, confirming his belief that there was probably no other alternative to research evaluation than to introduce quantitative criteria, such as the recruitment rates for PhD students.

When he became a minister in 1997, Claude Allègre received unconditional support from Pierre-Gilles, whom he occasionally cited to lend weight to his views, referring to the Nobel Prize in interviews with journalists. *"Listen to Pierre-Gilles de Gennes or Georges Charpak — both of them Nobel laureates in physics: they say what I'm saying"*, on the subject of the teaching in lycées.[12] Similarly, he quoted Pierre-Gilles de Gennes in arguing for his bill on innovation and research in parliament in June 1999 (a bill allowing researchers to create start-ups and file patents), though somewhat tying himself in knots on the details. *"When Pierre-Gilles de Gennes studies how a droplet of oil moves on a metal, he not only demonstrates a fundamental phenomenon of polymer physics, but also resolves a very difficult problem of engine lubrication."*

However, month on month, discontent with the minister amongst teachers and researchers grew. *"Claude Allègre's desire to knock everything down and rebuild it in his own way is, I think, understood neither by most of the scientific community, nor by his colleagues in Europe, let alone in the USA or Japan"*, wrote a well-known physicist to Pierre-Gilles in 2000. Nevertheless, the latter maintained his unflinching support for the minister. At the height of the discontent, when teachers were

---

[12] *Le Monde*, May 4, 2003.

on the streets, he even wrote on his behalf in a leading article in *Le Monde* in February 1999: "*We are at a watershed, when the minister's (welcome) frankness and (notorious) clumsiness have provoked violent reactions, aggravated by the circulation of an extraordinary number of false rumours. I am convinced that, on every educational issue, the teachers and the minister share essentially the same ideas. For the first time for a very long time, things can change: I am still hopeful.*"[13]

In March, however, Claude Allègre was finally obliged to resign. "*Overall, as regards university reform, it's a failure*", mourned Pierre-Gilles de Gennes. And this failure confirmed his belief that he would be no more successful than anyone else in loosening the rusty bolts of the French university system. "*Even Hubert Curien* (Author's note: Minister of research from 1984 to 1986, then from 1988 to 1993) *got nowhere, despite trying to make incremental changes*", he recorded. He threw in the towel. "*I have too often had the impression of not being heard (by ministers) to hope for better in the future*", he wrote to the head of the scientific and technical section at the Ministry of Research and Industry, refusing his proposal to join a think tank. "*Successive ministers are largely aware of the gravity of the situation, but the idea of a big university cleanup frightens them. As soon as university reform is undertaken, the teachers send the students onto the streets. No government is prepared to stand up to them. So the academic oligarchy persists, the teachers maintain their comfortable lives in deteriorating faculties, to the point that the universities have become a game of smoke and mirrors. Families think that their children are on the primrose path once they get into university, while it's not true*", he objected. "*We need revolution more than reform. But I can't see any government — right or left — that would be bold enough to launch it*", he mourned.[14]

Nonetheless, he did not withdraw from the debate. At the end of 2002, research was in crisis. Tired of funding cuts, researchers created the "*Sauvons la recherche*" [save research] collective. On October 16, Pierre-Gilles met the Minister Claudie Haigneré at the Palais de la découverte for a roundtable on the topic of young people and

---

[13] *Le Monde*, February 26, 1999.
[14] *Commentaire*, 108, 2004.

research, to talk about the lack of job prospects in France, which had become so severe that increasing numbers of French postdocs were remaining abroad and fewer and fewer students going into research. Following this, he was invited to meet the minister again, and explained to her the urgent need to create positions at the CNRS and at INSERM, the two spearheads of French research (though representing only 20% of the national headcount). *"If you provide posts now, it will not cost a huge amount. If you wait, the universities will mobilise, which will complicate your task"*, he emphasised. It was a lengthy meeting. *"The impression I retained from that long discussion was that the Minister and her office chief understood me very well. They probably tried to put the message across, but it didn't work"*, he regretted. So as Pierre-Gilles had predicted, the demands spread to the recruitment of university researchers. It was at this point that he backed off from the movement, refusing to support an increase in staff recruitment in universities in their current state. *"I then tiptoed away from the whole issue without regret"*, he said. He nevertheless put his name to the appeal by scientists against the 2003 Research budget, considered to be the *"worst possible budget for research and higher education"*, which attracted 5000 signatures: never had a scientist-led movement achieved such an impact.

Big Science

His other hobbyhorse was big science, what the French call TGE, to which he was violently opposed, arguing that these *"gigantic projects"* swallow up to 20% of the total research budget. *"In lean times like those we face today, it would be better to sacrifice big infrastructures to maintain recruitment among young scientists"*, he explained. On this subject too, he was a voice in the wilderness, despite his Nobel prize. For example, for years he fought France's *Soleil* synchrotron project. The story began when a number of physicists proposed replacing the old LURE synchrotron at Orsay with a new one. This was no minor undertaking: synchrotrons are huge rings (several hundreds of metres in circumference) in which electrons circulate at near-light speed, emitting x-rays, like the dust thrown out from the wheels in

Roman chariot races. These x-rays are then used to explore matter. When the project for the new Soleil synchrotron was drawn up, it immediately triggered a ferocious struggle between physicists, but the balance of forces was unequal. On one side were Pierre-Gilles and a handful of scientists, on the other the vast majority of the physics community. However Claude Allègre shared the Nobel laureate's point of view and, as Minister, rejected *Soleil* in 1999. The physicists backing the project refused to accept defeat, and referred the matter to the Parliamentary Office for the Evaluation of Scientific and Technological Choices, which organised a debate in 2000, at which Pierre-Gilles spoke for the objectors, after Georges Charpak, Jacques Friedel and many others had spoken in favour of the project. *"Only P.-G. de Gennes expressed great scepticism on the utility of megascience projects for research, explicitly calling for a substantial reduction in their share of the research budget"*, states the report by the SFP (French Physics Society), continuing: *"A large part of the discussion focused on the importance of synchrotron radiation in structural biology, and on the importance of that discipline itself, crucial for some, 'secondary' in the view of P.-G. de Gennes."* The Parliamentary Office concluded that *"the immediate construction in France of a national source of synchrotron radiation is an imperative for our country's scientific, educational and industrial dynamism"*, and *"should begin without further delay"*.

Pierre-Gilles continued to maintain that *"the scientific contribution of big machines is zero."* One major physicist tried to make him see reason. *"Your positions, because they are too extreme, are particularly dangerous, given your reputation and above all your influence on our Minister"*, he wrote to him in 2000. *"However, to contemptuously sweep aside everything that is not (yet) of Nobel level in physics and chemistry, or indeed to tar all the big machines with the same brush, shows above all your ignorance of the subject (...). Have you sat on the selection committees at ESRF or LURE?*[15] *No, of course not. How many times have you visited them or read their activity reports? (...) What, then, gives you the right to pontificate as you do?"* Pierre-Gilles stuck to his guns,

---

[15] LURE — *Electromagnetic Radiation Utilisation Laboratory* at Orsay. ESRF — *European Synchrotron Radiation Facility* is a European synchrotron in Grenoble.

reiterating that *"no important discovery has been made using a synchrotron."* *"What I see today is how hard it is for a postgraduate student to get a grant in biology. And after the PhD, to find a job. Yet biology is clearly the discipline that is on the rise at the moment. I prefer to support biology by creating young research teams than by the lines on an accelerator"*, he explained in a letter. His positions made him many enemies. One physician friend wrote to him: *"On occasion, at dinners, I have heard rather aggressive comments on the opinions you have expressed in various cases, stating your scepticism about big science projects. I would like to say that I agree with you, being convinced that these megaprojects reflect a mixture of political cynicism and economic muscle."* He replied: *"Thank you very much for your words. I am fighting a combination of conservatism and lobbying — but for the moment, they are winning."* And win they did. The *Soleil* synchrotron was inaugurated in 2006 at Saclay.

He was equally persistent in his opposition to the ITER (*International thermonuclear experimental reactor*) project, a prototype reactor designed to demonstrate the feasibility of thermonuclear fusion (A13-2). This time, he focused not only on the cost, but also on the ineptitude of the attempt, even suggesting that it presented risks that had not been properly evaluated. *"To believe that superconducting coils* (Author's note: necessary in this type of reactor), *(...) subjected to fast neutron streams comparable to those of an H-bomb, will have the capacity to survive throughout the entire lifespan of such a reactor (ten to twenty years), seems to me foolish. (...) If ever the superconductivity fails, because of the joule effect, the huge currents flowing in wires as thick as a human hair would instantly release a violent blast of heat. A superconducting coil is a potential bomb"*, he explained.[16] Although, this time, his point of view was shared by more scientists, he would have no more impact than before: ITER, run by a global consortium and supervised by the AEC in France, is currently being built in the Bouches-du-Rhône region, at Cadarache.

---

[16] Chantal Houzelle, "Research: Warning by a Nobel prizewinner, Pierre-Gilles de Gennes, Nobel Prize for physics 1991", *Les Échos*, January 12, 2006.

Hot and Cold on Europe

Pierre-Gilles had nevertheless moved heaven and earth to put across his point of view, writing as early as 1997 to Édith Cresson, a member of the European Commission, to explain why he thought the ITER project was an aberration, but in vain. Brussels had supported the reactor *"for reasons of political image"*, in his opinion. *"A serious error."* This was not the first time he had opposed Europe's policy on research. *"The [European] research agency is poor in its selection of projects; it seems to have little understanding of the scientific community. Proposals that involve a cocktail of attractive countries (from North and South) are easily accepted, with no critical scientific assessment. My personal feeling is that the ERA* (European Research Area) *always supports fields of science that have been around for a while (e.g. solid-state physics), and is insufficiently active in backing newer sectors (e.g. granular matter). But I may be wrong"*, he wrote to the European Commissioner responsible for conducting a survey on the European Community's science policy. His remarks having been ignored, two months later he followed up with a sharp letter: *"You are asking me to rubberstamp the results of your questionnaire. I have to say that I do not find many traces of the serious criticisms that I expressed on Europe's management. In conclusion, therefore, I would prefer you to remove my name from the survey completely."* His opposition to Brussels' research policy was so radical that he signed a protest document against the 6th European programme in 2001. However, he was not against the development of Europe, since he signed the appeal launched by the biologist Étienne-Émile Baulieu in 2005, which invited *"all involved in research in France to say yes to the constitutional treaty (...); because we believe that it is through excellence in research and innovation that Europe will be able to tackle the big economic and social challenges of the 21st century."* He said yes to the treaty, but a firm no to the policy.

Some physicists dream of nothing but winning the Nobel Prize. Pierre-Gilles was not one of them, by his own admission. Asked by a journalist in 2001 about mistakes in attribution of the prize, he wrote back: *"The most flagrant mistake is probably my own prize. Charles Franck, Jacques Friedel and Anatole Abragam should have been chosen ahead of me."* He made something of a joke of it: *"Given that there are around a million physicists in the world, each of whom works for around thirty years, and that the Nobel Prize for physics is often divided by three (...), it is 100 times easier to be honoured in Stockholm than to win the lottery."*[17] He also declared that the prize, as an individual award to scientists, was no longer in tune with the reality of research, which is now irrevocably based on teamwork, reiterating that he would have achieved nothing without his collaborators. Nevertheless, in response to a journalist who asked him whether the Nobel Prize should become a team award, he recommended staying with the status quo and instead creating new prizes: the Nobel Prize remains the ultimate symbol of scientific recognition.

---

[17] *L'Est républicain*, December 18, 1991.

Chapter 14

Life Goes On

The Nobel Prize did not change Pierre-Gilles de Gennes but it changed his life. He assumed that the excitement generated by the prize would die down after a few weeks, but it never really diminished. *"The tide of letters surged again after each TV appearance"*, he commented. True, his every word and deed were no longer reported in the press as they had been in the early days after the announcement — one newspaper even reported the time he trapped his fingers in the car door in October 1991 — but he remained very much in demand. *"The Nobel Prize means an average of an extra hour's work a day for the rest of one's life"*, he reckoned.

With all the media hype and the school tours, he spent less time in the lab. *"For around a year, we didn't see much of him"*, concedes Madeleine Veyssié, director of the lab at the time. *"We consoled ourselves by watching him on television"*, she jokes. Nonetheless, he never stopped working. *"Even immediately after the Nobel, he was still borrowing books. We used to go and fetch them from different libraries around Paris. We would give him one book a day, and he would usually return it the next day, having read it in 24 hours!"*, recalls Yvette Heffer, his secretary at ESPCI. In 1992, when Liliane Léger and Hubert Hervet, from his lab at the Collège, asked him to look at their experimental results, he followed them with undiminished interest. They had measured the

263

speed of flow of a liquid polymer on a surface covered in "anchored chains" like a deep pile carpet. With ordinary liquids, for example water flowing in a pipe, the speed at the wall is always zero. With the polymer, it was different: the speed at the wall rose above zero as the flow accelerated. The following Monday, Pierre-Gilles brought a model which explained that the chains attached to the wall stretch and untangle to form a sort of slide, on which the other chains can move, with the result that their speed at the wall is no longer zero. This model led to a redefinition of all polymer friction mechanisms. He also invented a "molecular drill" with Maya Dvolaitzky and Marie-Alice Guedeau-Boudeville (Booboo) in 1993: this was a porous grain which attached to a lipid bilayer and could drill a microscopic hole in it. He continued his investigations on adhesion, wetting, etc. In all, he published 11, 13 and 18 articles respectively in 1992, 1993 and 1994 (compared with a global average rate of 14 a year).

Nor was his insatiable curiosity in any way diminished. He could go to any lecture, even on palaeontology, and would emerge with new ideas. *"Yesterday (…), you mentioned (I believe) that the manufacture of two-sided stone tools (bifaces) spread in only 50,000 years. This seemed like a short time, but on reflection I think that it is right. The argument I am thinking of is based on the spread of information. It can be presented as follows: a group of individuals A, familiar with the biface, covers a range of 1–10 km on its daily travels. If it meets a group B, unfamiliar with the biface, a specimen of which is either traded or taken, I imagine that B would study the object and try to reproduce it. Naïvely, I would say that after a learning time of around 3 months, it would succeed. Knowledge spreads (…) with what we call a diffusion coefficient of $D = F/t \sim 1\ km^2$ per day. (…) Such a process takes a time $T = L^2/2D$. If $T = 50,000$ years, $L \approx 6000$ km. This means that, starting from a fairly central point in Europe + Asia + Africa, we completely cover this block. This argument is probably simplistic, and I am perhaps seriously out in my estimate of the value of learning time t. But the calculation is so simple overall that I couldn't resist the temptation of suggesting it"*, he wrote to Yves Coppens (in 2001). He remained on the lookout for new subjects, for example attending a conference on drug delivery methods in 1993. In 1995, he began studying "artificial muscles", gels capable of contracting when subjected to an electrical discharge, in collaboration with Philippe Auroy of the Institut Curie. *"I spent years on this subject, but I never got anything very interesting out of it"*, he regretted. In 2002, he set foot — for the first time — in Russia, to attend a symposium on polymers in St Petersburg, where a physicist colleague presented him with the following problem. If two species of incompatible polymers are mixed in identical proportions, on the same chain, then the polymers form micelles (small balls a few tenths of a micrometer across). But what happens when the proportions are chosen randomly? More mathematical than physical, the question gnawed away at Pierre-Gilles, who left the conference to stroll around the city and think. In the course of his wanderings, he came to a wide avenue, sat down on a bench and let his thoughts roam. *"I was ruminating"*, he said. A sailor, completely drunk, greeted him and sat down next to him. Pierre-Gilles answered with a few words of Russian he remembered from his military service on board the *Richelieu* in 1959, while still thinking about the problem. A solution came to him. He saw how to use the fluctuations to predict the demixation (separation of a liquid into several phases). He leapt to his feet, leaving the sailor to his blatherings, and returned to the hotel at top speed to put down his ideas in black-and-white. Beyond these different subjects, however, the next big topic he tackled after his Nobel prize was the physics of sand.

Shifting Sands

Pierre-Gilles had long been interested in the physics of sand. *"My attempts to understand sand heaps were too crude and I discarded them (but send me your publications)"*, he wrote to the Japanese physicist Yoshihiro Taguchi in 1990. It has to be said that sand contained many mysteries, each more intriguing than the last. For example, it flows through the fingers like a liquid, whilst forming an unquestionably solid heap (load-bearing), yet its surface grains blow away in the wind and collide with each other like the particles of a gas. It is neither liquid, solid, nor gas, but a bit of all three, depending on the circumstances. This means that hydrodynamic, mechanical or statistical approaches gave a correct description of situations where the sand could respectively be likened to a liquid, a solid or a gas, but became inappropriate otherwise: there was no unified theory of the dynamics of a granular medium. Theorists like the British physicist Sam Edwards thought that sand obeys its own thermodynamic laws, but were unable to determine how the energy produced by friction and collisions between the grains is dissipated. Here was a subject that could not but attract Pierre-Gilles, especially as granular materials (gravel, cereals, sugar, coffee, pills, etc.) are involved in numerous industrial applications!

The theorist began by thinking about the way that grains settle in a silo during filling, which usually results in the formation of cones (or arches) that obstruct the flow from the exit funnel. In 1995, he published a first study on the distribution of pressure in a silo during the filling process (A14-1). To take the subject further, he decided to devote his next course at the Collège de France to granular materials. In 1997, he asked Jacques Duran, director of studies at ESPCI and a specialist on the physics of sand, to lend him his course notes, and picked out a stack of articles before leaving for the summer vacation. And by September, as usual, his course was ready! While preparing his lectures, he also — once again, as usual — reflected on a number of outstanding problems, giving special attention to the so-called BCRE model of grain avalanches (named after its authors, J.-P. Bouchaud, M. E. Cates, J. Ravi Prakash and S. F. Edwards) (A14-2). Pierre-Gilles introduced a new, intermediate, so-called neutral angle, between the already defined angle of repose (where all the grains are motionless, but on the verge of slipping) and the angle of motion (where the grains start to move), which significantly simplified the equations (A14-3). This work culminated in the "Bouchaud-de Gennes" model, used in the study of different types of avalanche. For example, in describing the "flow" of an avalanche of sand on a dune, it shows that we end up with the same equations when we start from the descriptive BCRE model as when we rely on the fundamental principles of hydrodynamics. He thus took a first step toward unifying different approaches. However, it fails to elucidate what triggers another avalanche: is it one grain in disequilibrium which falls on another and starts an avalanche by a domino effect (localised phenomenon)? Or do certain unstable zones start sliding *en masse* (delocalised phenomenon)?

In any case, Pierre-Gilles' take on the physics of granular materials had the impact of a jack-in-the-box. His simplifications of the BCRE model aroused a furore with the authors of that model. *"It is rather as if he had cut off their baby's arms — to clarify the model — and that, despite everything, the baby still seemed capable of walking and developing. They couldn't get over it...",* recalls Thomas Boutreux, the theorist's doctoral student at the time. After initially contesting the radical hypotheses put forward by Pierre-Gilles de Gennes, they eventually accepted his point of view.

Then Pierre-Gilles turned his attention to mixtures of grains of different sizes (or with different surface properties) and explained certain segregation phenomena (the spontaneous separation of different grains under the influence of movement or vibration) (A14-4), but still wanted to go further. He clarified concepts and introduced effective new models, but was no more successful than anyone else in identifying universal laws. Aware of this, he moved away from the subject, despite his continuing interest. *"I remember a conference about the desert — very interdisciplinary (there were even poets there) — in Tunisia. We went into the desert to look at the stars and watch the sand avalanches on the dunes. I found myself in conversation with Jean Daniel, who had given a talk on Albert Camus. I really liked the field of granular materials"*, he admitted. When he left it, he did not — as he sometimes had before (in superconductivity and liquid crystals) — leave behind an experimental team (he had done no laboratory experiments on the subject), but just a few students, though in this case they did not feel abandoned. *"It is as if he had killed a bison, eaten the best parts and gone off, but there were plenty of "leftovers" for the rest of us! And then, we could always go and talk to him"*, reports Thomas Boutreux. Overall, Pierre-Gilles felt that he had not *"contributed much to the physics of sand"*, apart from an article where he intervened in a dispute between specialists in fluid mechanics and physicists, for once proving that the traditional school of fluid mechanics was right. He saw the period 1995–1999 as one of his least productive.

"There was a lot going on, but also a lot that never really went anywhere", he judged.

Making the Desert Bloom

In 1998, he was contacted by Pierre Rognon, emeritus professor at Pierre et Marie Curie University in Paris, who was looking for physicists to study the advance of sand dunes and to combat desertification. The theorist directed him to other researchers, but promised him: *"You have my support."* The two men stayed in contact. In the early 2000s, Pierre Rognon approached him again for help with the following problem. When it rained in Niger, the water would sink into the ground too quickly to be captured by plants. This could be resolved by using hydrophilic polymers to retain the water on the surface. However, existing water-retaining gels were too expensive for Niger. Pierre-Gilles became interested and wanted to launch a research programme with European funds. He contacted a colleague familiar with the ins-and-outs of funding in Brussels. *"It would need a fairly large-scale operation involving research laboratories and chemical firms, to achieve any significant outcome. Is there a European channel for subsidising this type of initiative?"*, he asked in 2002. Then, on the advice of a manager at *Rhodia*, he approached the firm *Procter & Gamble*. *"I've had a crazy idea: to recover used diapers, break them up, wash them and send the material to Niger. (...) We think that this is a project that could interest Procter & Gamble. They could launch a public advertising campaign: "Help us to recycle your used diapers to support the Third World"*, he wrote in 2004 to a female friend at the company. *Procter & Gamble* studied the idea, but rejected it after a few months to the physicist's great regret.

Strategic Network

Alongside his research on granular materials, he was exploring subjects as varied as the behaviour of a colloid grafted into a mixture of water and oil, the propagation of viscous droplets on a non-viscous liquid or the injection of starshaped polymers into a nanopore! He was never short of ideas, keeping abreast of developments through scrupulous reading and his huge network of fellow physicists around the world (even more after the Nobel Prize). Thanks to them, he received and processed enormous quantities of information. From it he drew more ideas than he had time to consider. He would select the most unusual or the most promising, distributing the rest. He cultivated his contacts — an idea for a subject, a friendly comment — and the grateful recipients would return the compliment, alerting him to things that had aroused their interest. In this way, nothing passed him by.

With his extensive network, he could have built himself an empire in the world of research, so ready were physicists to seek his opinion (whom would he choose for this prize or that position?) and to please him... However, he responded honestly to each question, without necessarily advancing the cause of his friends. When, in 2000, a director of the CNRS asked him which physicists had the qualities for high positions in the organisation, he answered: *"I tried to find names of dynamic physicists who could assess and develop both mature sectors of physics and growing sectors"* and recommended four researchers, describing them precisely and objectively: the first was an *"exceptional motivator"*, the second *"extremely energetic and efficient and effective"*, the third *"has very good judgement — but perhaps slightly distorted by the 'ENS' spirit"* and the last was *"younger than the others"*, which made him wonder whether it would not *"do him a disservice to distract him from research at his age"*. Similarly, questioned on a new director for the Institute of Surface and Interface Chemistry in Mulhouse, he suggested several names, including that of a team leader in his lab at the Collège. *"It would be a loss for us, but (...) we can't find him a good position in Paris"*, he wrote in 2000. A word from him could open many doors, but he did not exploit this to play favourites. He could also have increased his influence by sitting on multiple juries and handing out prizes where he wanted, but only rarely agreed to join selection committees, so reluctant was he to waste time studying applications. Nonetheless he was loyal to *l'Oréal's Women in Science* prize, awarded to five female researchers of world stature. He was also committed to the Institut de France jury, which selected deserving teachers to receive bonuses and time out of teaching. *"I've got 20 reports to write, which takes time..."*, he wrote to his friend Phil Pincus in 2002.

On the other hand, he was ready to use his influence to correct situations that he found unacceptable. As far back as 1973, for example, drawing on his reputation in liquid crystals, he wrote to the editor-in-chief of *Physical Review Letters* (*PRL*): *"I would like to express some concerns regarding your selection of short articles on liquid crystals. It is my impression that, in this fast changing field, you have not yet put together a network of peer reviewers of the quality and effectiveness to be found in other fields."* And he gives four examples of articles which, in his opinion, did not deserve to be published — *"I have searched in vain for a single interesting point in this article"*, he comments baldly on one of them — and goes on to cite two unfairly rejected French articles. *"The authors have shown me the reports of the two referees. I cannot help thinking that these reports actually come from two rival teams that are conducting measurements on the same subject."* In response, he received the list of the 14 principal referees in the field of liquid crystals, with this proposal from the editor: *"You could help us by suggesting which names should be added or removed."* Pierre-Gilles took up the challenge and answered by crossing out two names on the list — *"as I see it, [they] represent a slightly obsolete style of research in physical chemistry"*, he explains — and by suggesting three new names (including two European scientists). He got his way.

All in all, he was extremely alert with regard to fairness in the publication of articles by French physicists and the recognition of their work. In 1990, he denounced the publication of an article on high-temperature superconductivity in *Physical Review Letters*, which described similar results to those obtained three years earlier by a physicist working at ESPCI, without citing this earlier work. *"It may seem a minor matter, but it is not (...). I find that certain American scientists systematically try to bury the contributions of competing teams. I personally experienced an attempt of this kind a year ago, when one of my old articles on smectics was used in a PRL article, even using the same notation, without being cited"*, he complained. It was a serious accusation, especially coming from a Nobel laureate. However, the recognition of individual work — and particularly his own — was a matter he took so much to heart that, despite his equable temperament, he was capable of flying into a rage if he discovered that he had not been properly acknowledged.

The same spirit of fairness prompted him to write to the great English physicist, Sam Edwards, in 1976, to express his discontent when he learned that Edwards had organised a British-only conference, whereas two years earlier he, Pierre-Gilles, has set up a Franco-British meeting on liquid crystals. *"I was a little surprised to learn (by accident) that a liquid crystals forum was held in Leeds on April 5–6, 1976 (...), without any French involvement. I can understand the need for purely national conferences, for example when a new field emerges (...). But that is not the case here. Coordination between the programmes in England and France (on questions such as deuterium resonance or neutron scattering) would in fact have prevented pointless duplication. (...) I would like to say that it is not a personal matter, since I now hardly work on liquid crystals any more. (...) But this gives me more freedom to express my disappointment."*

Other than in such cases, he did not exploit or try to extend his influence. He kept the institutions at arm's length and rejected a political career. True, he played the media game, but he refused all compromise, which prompted him to leave the scientific committee of *La Recherche*. *"Your journal,* La Recherche, *has been highly successful. Personally, however, I find myself not at all in agreement with many of its positions: its antinuclear stance, its attitude in the history of science and, more generally, with a certain arrogance in its journalists"*, he explained in a 1988 letter to the chief editor. In 1995, he also declined a proposal to join the Hubert Beuve-Méry association (made up of readers of the *Le Monde* newspaper), giving three reasons: *"a) a certain cliquey arrogance, which does not always encourage dialogue. I remember a time (long ago) when the paper asked me for information on the grandes écoles: I answered more broadly with a fairly severe critique of the system, which was never published. b) an ambiguous attitude to pseudoscience (e.g. recent articles on the memory of water). c) Somewhat crude environmental positions, which fail to reflect the real problems (...)."* Overall, he was mistrustful of the media. *"I won the Nobel prize for my contribution to the understanding of soft matter, in which extremely small effects have incredible consequences. I have discovered that the media too are capable of transforming public opinion. Our media-obsessed society is a "soft" society."*[1]

Ultimately, he chose to maintain his independence rather than use his influence, no doubt too aware of the limits it would anyway have, having had his fingers burned in a few unsuccessful attempts at reform. So he did not try to build an empire, often taking care to stay at arm's length from power struggles in the scientific community, where the inner circle of his collaborators (nicknamed the "deGennerie" by the acerbic wits who were not members), which was ultimately quite small, had little influence: the chapel of soft matter remained minuscule in comparison with the great cathedrals of physics, as unassailable as their mediaeval architectural counterparts.

Pierre-Gilles as far as possible avoided involvement in the intrigues of the scientific world, which led some to describe him as *"someone who tends to wash his hands of problems, both to protect himself and not to waste time."* However, he was no recluse and had no hesitation in supporting his colleagues when the situation warranted. *"He's one of those classy people who helps others without advertising the fact"*, comments one physicist. Indeed, he intervened more frequently than was apparent — when the situation demanded. For example, he wrote to defend a researcher accused of having delayed an HIV test to promote the interests of *Diagnostic-Pasteur*: *"This accusation seems to me psychologically unsustainable"*; to support the polymer physicist, Jacques des Cloizeaux, "menaced" with retirement: *"We're talking here about a major theorist. His work (...) is (...) historic"*; to defend a physicist friend who was in conflict with a researcher in his university: *"I can assure you that his work is exceptional."* He even wrote to propose that the Légion d'Honneur should be awarded to his old NSE classmate, Hubert Gié, knowing that it would give him pleasure. *"He had become a senior schools inspector, a pillar of science teaching in France. We remained very close"*, he said. On occasion, he would write

---

[1] *Le Point*, December 14, 1991.

spontaneously, for example to help a colleague who was hesitating between several positions: *"He is an exceptional researcher and I'm worried about him. Can you talk to him?"*, he asked a fellow physicist.

He was no longer as acerbic as he could be in his early days, both with his colleagues and with scientists of the greatest stature. Had he not let fly at Philippe Nozières, on discovering his course programme at the Collège de France: *"Surely you're not going to redo a Bruhat?"*, Georges Bruhat's book being the epitome of classical physics. He could still sometimes be blunt in his remarks, but always with experienced and (over) self-confident researchers. *"Your work shows a wide range of knowledge — but in my life, I have seen too many theorists trying to impose fashionable concepts on a real phenomenon, when it is the opposite you should do: start from the bottom up (...). Conclusion: I am totally sceptical, but of course, I may be wrong"*, he wrote sharply to one of them in 2001.

"He mellowed with time", confirms one of his early female collaborators. He cared more about his colleagues, at times even being attentive. For example, learning that one of the laboratory's former students, a post-doc in California, was thinking of giving up research, he immediately wrote: *"I would like to give you my immediate reaction. I too, as a young student with Aigrain, and then as a novice researcher at Saclay, thought several times that my experiments would never work and that research was not a path for me. (...) It is a normal phase. But in your case, [we] can firmly maintain that you have what it takes to be a very good researcher. You have to think about all this calmly (...). Take your time, enjoy California and your family, and everything will gradually fall into place"*, he advised. When one of his PhD students decided to quit research, he was disappointed, but respected his decision and wrote to give him contact details for senior figures in industry. He had not bothered with such things at the start of his career.

He would also sometimes intervene in favour of people he knew only from a distance (but who had made a good impression on him). For example, he wrote to Simone Veil, Minister of Social, Health and Urban Affairs, to request citizenship for an African teacher whom he had known as honorary president of the Philotechnical Society. *"Please forgive me, Madam Minister, for disturbing you with this particular case"*, he began before explaining that this young maths teacher had *"shown exceptional energy"* in giving remedial lessons to inner-city children. He also supported the application of a secretary with whom he had worked: *"I was impressed by her qualities — in terms of efficiency and accuracy — but also in more human terms: by her good humour and optimism, she creates a remarkable working atmosphere around her. I am therefore very happy to recommend her."* The parties concerned would never know anything of these interventions.

However, he was no samaritan. He never lost sight of his priorities. *"He wasn't a man to stop and chat. The atmosphere was all work, but he could sometimes let slip a friendly remark. "You've typed so many articles that you could write them in my place", he joked to me one day"*, recalls Yvette Heffer, his secretary at ESPCI. And his priorities were scientific. They say you can't teach an old dog new tricks, and from the 1990s he began to grumble about the new fashions he saw emerging. For example, he refused to take part in a conference on fractals: *"This has become a very popular topic, which attracts far too much media hype."* Similarly, when asked his opinion on the creation of a Complexity Institute, at a time when it was a very fashionable subject, he did not beat around the bush: *"1) I find the term pretentious and ill-defined. 2) I am not at all convinced of the existence of profound links between the [different] domains in your list. For example, I do not believe in the overlap between biochemistry and chaos, or between new materials and economics, etc. 3) I think it would be extremely difficult for a student to try to learn the basics of all these fields (...). The result is likely to be a superficial grasp (...)."* He had no hesitation in expressing his reservations to senior scientific figures, such as the chief executive of the CEA in 2003: *"I fear (...) that initiatives will be maintained in fashionable programmes (e.g. nanotechnologies) which will not necessarily have the anticipated scientific and economic impact (I was already giving courses on nano-objects at the Collège de France ten years ago). I implore you to revise this plan."*

Uproar among Physicists

Pierre-Gilles had strong ideas and often expressed them bluntly, for example describing television and information technology as *"instruments of dumbing down"*, sometimes letting his love of a colour-ful expression get the better of him. In fact, the reality was more subtle. *"He hated computers, but he said to me several times: "you really must teach me to use one"*, chuckled Yvette Heffer. In research, he certainly never ran computer simulations himself, but he did not denigrate those performed in his lab, which even provided inspiration for some of his results. As regards television, his position was somewhat the same. True, he did not have one at home in Orsay, but he was happy to spend TV evenings with friends with Annie and their children. In any case, his bluntness generated controversy. The biggest was confined to the research community, but brought down the wrath of all, or almost all his colleagues, on his head. It all started with an article in *Sciences et Avenir* in 2001. *"We need to acceler-ate the slow death of exhausted fields such as nuclear physics, solid-state physics or quantum physics"*, he wrote. A call to arms: the physicists in question (of whom there were many!) reacted vehemently. Jacques Friedel wrote him a *"sorrowful"* letter. Jean-Paul Hurault, one of his former pupils, sent him an official letter of protest, in his capacity as President of the French Society of Physics. The Academy of Sciences also remonstrated with him. *"I also received messages from PhD students alarmed by my words. I was upset"*, he confessed. He answered them all. *"The text in* Sciences et Avenir *is not at all accurate (I had said "limit [the development]" not "accelerate the slow death"), but I take respon-sibility for the general tenor"*, he wrote to Jacques Friedel. But the damage was done. Pierre-Gilles claimed that, when proofreading the article for publication, he had asked for the sentence to be changed, but this had not happened. In any case, it was not the first time that the soft matter physicist had, in one way or another, expressed an opinion of this kind. For him, the age of big discover-ies in physics was over. In his inaugural class at the Collège de France in 1971, he had observed that *"a large fraction [of the] mysteries [of solid-state physics] [had] melted away"* and that he was *"convinced that a profound (...) change [was] needed in solid-state physicists"*. Moreover, some ten years later, giving the opening speech at an exhibition as President of the French Society of Physics, he had sparked an early skirmish by declaring: *"There is too much money spent on low-temperature*

research, which is no longer very productive." The physicists in the room were furious, and took pains to remind him when high-temperature superconductivity and so-called "giant magnetoresistance" (which won the Frenchman Français Albert Fert the Nobel Prize for physics in 2007) were discovered one after the other in 1986 and 1988. The theorist simply shrugged. These achievements in solid-state physics did nothing to falsify his assessment: the next big scientific break-throughs would not be in physics (but he had never said that there was nothing left to discover).

Since the late 1960s, he had felt that the future of research was in biology, not in physics. *"When I used to open* Physical Review Letters *in 1960, I would find a revolutionary idea in every issue. These days, I come across two or three ideas a year, although the journal is five times thicker"*, he commented. *"Physics is a science of simple systems — the simpler the better. It is entirely different from biology: a living system always constitutes a long, convoluted and historical construction. Biology has a huge future of discoveries for the 21st century. Physics has a more modest schedule of inventions: building new objects on the basis of well-known ideas, and making them useful"*, he wrote in 2001.

Would he continue to promote biological research in this way and constantly boast of its prodigious future, without venturing there himself? This is the question he asked himself as the time to retire approached, both at ESPCI and at the Collège de France. In fact, he was readying himself for a strange retirement.

Chapter 15

"God Willing"

"It's exciting. A bit like going back to school...", exclaimed Pierre-Gilles de Gennes as he arrived at the Institut Curie.[1] Instead of retiring, he was making his biggest ever career move by starting new research... in biology.

He had long been interested in biology. He was curious about everything, but obsessively so about nature. He had developed the soul of a naturalist as a teenager, during his hikes in the Alps, and especially over the two years at the NSE, with its focus on the natural sciences: he had drawn dozens of digestive tracts or cross-sections of pistils. *"We had a very good grounding in the biology of the time. In particular, we knew a lot about cellular and hormonal biology. I could describe the successive stages that lead from the larva to the butterfly and the hormonal mechanisms behind these changes. We also knew about the natural environment, animals and plants, which I was greatly interested in"*, he recorded. His mind went back, too, to his oceanography course in Banyuls-sur-Mer, collecting seaweed and catching cave-dwelling fish.

In 1951, when he entered the École Normale Supérieure, he nevertheless had no hesitation about going into physics research — *"Physics*

---

[1] *L'Humanité Hebdo,* December 6, 2003.

seemed a more precise science, which I liked", he explained. Over the early years, he did not deviate from his choice, but from the mid-1960s began to allow himself ventures into biology, as he would regularly in the course of his career. He wrote his first article relating to biology in 1967, on DNA denaturation (the separation of the two strands of deoxyribonucleic acid). The following year, he wrote two more on the energies involved in the different configurations formed by biological structures such as proteins, adopting a highly theoretical approach based on an analogy with... quantum particles. He also published a study in 1973 on the rupture of red blood cell membranes when subjected to mechanical stresses. Moving away from the field for some years, he returned in 1993 with several publications, for example bacterial swimming. He would sometimes extrapolate the results he obtained in physics to biology, for instance moving from the introduction of polymers into pores to the penetration of DNA into cells.

Although he only published episodically, he would always keep abreast of advances in biology. Initially, his information came mainly from his former naturalist friends from the École Normale Supérieure. *"My friends had not had a very inspiring education — the teaching was old-fashioned in the 1950s, and some French biologists even tried to exclude molecular biology* (Author's note: DNA was discovered in 1952) *— but they subsequently caught up. Maxime Guinnebault had studied kidney physiology with radioactive markers, which was his entry into modern science. Pierre Favard, aware that their generation was a little "behind", solved the problem by doing electron microscopy. We used to tell each other about our research"*, he recalled. Then, from the mid-60s onwards, the physicist wanted to learn more and attended Saturday morning lectures at the Institut Pasteur. His particular interest was progress in DNA sequencing. *"The team led by C. Reiss at Orsay has just demonstrated the effectiveness of the extraction method proposed by the theorist Azbel: it can identify 23 sequences of intermediate length and give their composition. Although a nonspecialist, I have been impressed by this result which seems to open up significant possibilities (...). My only contribution to this affair was to talk to Azbel last year, to put him in contact with Reiss in the autumn and to see the results appear six months later!"*, he wrote to a colleague in 1978. Molecular biology was not his only interest. Around the same time, he was receiving research reports from the... Monaco Oceanographic Institute! In the 1990s, he became fascinated by neurobiology, singing the praises of researchers who were trying to *"see neurons"* on Paul Amar's TV programme *20 h 00 Paris Première*. When in April 1998 he was invited with Annie to the lab in Banyuls-sur-mer, where he had not set foot since his youth, he agreed to return to talk about current research, not to reminisce about *"old times"*.

His knowledge of and interest in biology were well-known, to the point that biologists deliberately sought him out. In 1977, a researcher at the Institut Pasteur suggested a subject for him: *"In proteins subject to allosteric regulation, information transfer necessarily involves the interface between the subunits. In certain cases, the small bound molecule partially modifies this interface and it is the cumulative effect of the binding of two or three molecules that produces a decisive change in the conformation at the interface. (...) Would you be interested?"* In 1994, he was approached to join a committee to work on the publication of the minutes of the Biology Society (which he accepted) and, in 2000, to explore the future of research at the Institut Pasteur.

A Career in Biophysics

In the mid-1960s, particularly after his meeting with Charles Sadron, a polymer specialist interested in biopolymers, there was a strong temptation to move into biophysics research. *"Biological processes are physical and chemical mechanisms! It was therefore natural that, at an early age, Pierre-Gilles de Gennes should have been tempted to combine physics, chemistry and biology"*, reflects the physicist Julien Bok. However, he gave up this idea after dipping his toe in the water a few times (hence the articles in this field), without ever regretting it. *"I have often found that the careers in biophysics available to physicists — such as establishing the detailed structure of proteins — were useful, but not very imaginative"*, he explained.

However, his views on biophysics changed in the 1990s: *"I am fascinated by two things: a) the advances in different subjects, for example the operation of the inner ear, b) recent developments in bioengineering, in drug delivery and in biomedical studies. I am convinced that these sectors can and will improve our lives"*, he wrote in 2002. It was in the 1990s that he published another series of biology-related articles. *"[Biophysics] has become very fashionable among physicists. But some serious sorting needs to be done. There are good programmes for new instruments, for example microscopy in living systems (...). In my mind, the issue of biomedical research is much more important: it is much less effective in France than in the US, because the medical community here doesn't mix much with the other disciplines"*, he believed, going on to stress the importance of subjects *"halfway between physical chemistry and medicine, for example bone repair, etc."*

Despite this revival in interest, he did not take the step. *"Personally, I have never considered moving into biology, and I am still trying to do physics (on glass transition, artificial muscles, etc.)"*, he wrote to Jacques Friedel in January 2000. Indeed, at this time he was trying to understand the nature of glass. *"I have been interested in the abnormally low diffraction angle that Fisher observed in several systems. I am convinced that there are clusters (at all scales) and I am trying to understand their role... Difficult"*, he wrote to Phil Pincus.

Yet two years later, he was moving into the Institut Curie, a combined research centre and hospital dedicated to the fight against cancer, to do research on biological questions. What induced this U-turn? In truth, he had left the Collège de France under compulsion. Comfortable as he was there, he had no choice but to step down at the age of 70 — that was the rule: no one was allowed to grow old gracefully at the Collège. Out of the question for him to settle at ESPCI, since he was in the process of handing over the reins to Jacques Prost: how could his successor do his job with Pierre-Gilles de Gennes in the backseat? In addition, he had never done research there, only run it... So he started to look for somewhere else to go. Retire properly and stay at home? No way! He would have been a fish out of water. Fortunately, he was offered a position as adviser to the President of the Institut Curie, which suited him perfectly.

Indeed, the aim of the Institut Curie was to promote research at the interface between physics and cellular biology: was this not the direction he had always sought? So it was not surprising to find a good number of his former students working there, such as Jean-François Joanny, his first PhD student at the Collège, Francis Rondelez and... Françoise, who had been doing biophysics research there for years, alongside experimentalists. In short, he was on familiar ground. *"I had followed Edmond Bauer's classes in the next-door building, when I was at the ENS"*, he recalled.

In his first months at the Institut Curie, he continued research begun with Françoise on cell adhesion. Seeking to understand how one cell recognises another and binds to it, he began by considering the simplest case of a vesicle adhering to the surface of a lipid bilayer. He showed how, once the vesicle is in place, its initial surface tension influences the growth of the contact zone. However, he was not satisfied with his model, which he considered to be *"biologically unhelpful"*, as the mechanism is controlled by the surface tension of the vesicle, whereas, in a real cell, it depends on the deformation forces of the actin network (the network of contractile proteins). *"Overall, the role of theory in biophysics is very modest: we are handmaidens to the biochemists and physiologists. I even find that the biologists have often trumped the physicists in finding new techniques, for example separating DNA fragments by gel electrophoresis (a subject I have followed); it is they who created the method and devised the major techniques (...). Theory came in after the battle was won"*, he would write to a friend in 2004. He abandoned the subject the same year. *"Françoise was doing fine without me..."*, he acknowledged.

However, there was another reason for this change. He was deliberating on an ambitious — not to say extravagant — plan, which excited him far more: he had decided to start doing research in neuroscience. It was an idea that had secretly attracted him for a long time. *"To study the brain seriously, you have to work for at least three years, in the right place, as a student, to develop the dual culture. A good friend of mine, someone I like a lot, Charlie Bean, a physicist, did that: at the age of 50, he went to the Rockfeller Institute in New York to study the brains of cats. Those are the kind of people I respect. (...) Learning by reading books or listening to lectures, yes, I'd very much like to do it"*, he announced in an article in *Sciences et Avenir* in... 1991! But between wanting and doing, there was a big gap. Things clicked in 2003, when he started to read his daughter Claire's thesis on neuroscience. He didn't understand a word. *"You'll have to initiate me; it's really interesting"*, he said to her. He began learning neurophysiology, not only to decipher his daughter's manuscript, but also to start research in the field himself. He was aware that he was on treacherous ground. Numerous physicists before him, including some big names, had ventured into biological research, with varying degrees of success: Brian Josephson, Leon Cooper, Erwin Schrödinger... It could be a tricky transition at any time, let alone in his seventies! He did not repeat the mistake he made in 1979, following his work on flow in suspended particles. *"I had been trying to understand erosion mechanisms, for example how river basins form. I had constructed a small model of erosion, but had not done enough work on it. As a result, it was of no interest to the specialists"*, he acknowledged. So he worked flat-out, starting from scratch, like a 20-year-old student.

Memory Storage

He spent two years learning about the field, using his well-established method — bibliography, reading, identification of specialists, identification of problems — which led to his choice of subject, in this case "memory neurons". He asked himself a question that is easy to state — where and how is a memory stored? — but whose answer remains hidden in the meanders of the brain. The physicist turned apprentice neurobiologist began his search.

"If I smell a rose, in what form is the scent first stored in the brain?", he wondered. At the time, some neurobiologists thought that it was stored in a single cell, whereas others believed that it required numerous scattered cells. Pierre-Gilles reached an intermediate conclusion. According to him, three or four cells are enough to store an odour. He reached this surprising conclusion by considering the number of odours stored in the memory (around a million) and the wiring capacity of the brain. He also thought that there is a certain redundancy built into the storage process. For example, the scent of the rose is associated with different memories and stored in

several places. As a result, even if the primary "scent" memory is lost, it can reappear, stored elsewhere in another form. In this way he established a simplified model of olfactory memory, which shows how the pulse train from the olfactory bulb excites cells in the piriform cortex, which in turn transfers the information to a storage zone.

As was his habit (in physics), he submitted his model to colleagues specialising in neuroscience before sending it to a journal, this time apologising for his encroachment on their turf. *"I have been struck down by a disease typical of old scientists: becoming interested in the operation of the brain. My approach is crude and quite different from [yours]. But I have decided to send you this first attempt..."*, he wrote in march 2004 to the physicist Daniel Amit, another convert to neuroscience. When he finally published his model in 2005, it was favourably received (though not so far confirmed).

Pierre-Gilles considered his research on memory storage to be amongst the hardest of his career, not because of the calculations, which he described as *"elementary"*, but because of the completely different method he had to learn. *"In physics, we try to model a system using the simplest possible route. Biology is more like archaeology: when you excavate a town, you find a house that was initially built for a modest purpose, and was then enlarged and remodelled over many generations. Our brain is not a well-designed system; it is a cobbled together system"*, he wrote in 2004. He had worked hard to achieve this model, but it had also brought him *"great pleasure"*, proving, above all to himself, that he was able to take on the challenge. He did not stop at one article, publishing another study in March 2007 on neuron growth and the self-organisation of axons (extensions to the neurons).

Still Physics

Nonetheless, Pierre-Gilles did not abandon his primary field, since alongside this research he was tackling new subjects in physics. For example, an article sent to him by Manoj Chaudhury, from Lehigh University in the US, alerted him to a curious phenomenon. The article described the motion of a drop of water deposited on a medium subject to dissymmetric vibrations (i.e. shaken hard in one direction and gently in the other). Pierre-Gilles found the effect interesting, but did not agree with the author's interpretation. Thinking about it, he came up with a model that takes account of wetting hysteresis. *"And what would happen to a coin on a table that vibrated like this?"*, he wondered. He did the calculations, which showed that ... the coin moves forward. He then wondered what would happen if the coin was subjected to random vibrations. *"I had an inkling of the answer based on a scaling response."* He wanted to find someone else to do the calculations, but could not. *"So I did all the calculations myself, which was quite a job, but I succeeded"*, he boasted The result was as expected: the coin adopts a Brownian motion with dry friction (a random movement that includes friction). In the process, he studied the case of a drop deposited on a lattice subject to asymmetric vibration, in which capillarity must also be taken into account. *"So this business of the motion of droplets, which initially came from an article I had received, stimulated articles in at least three directions!"*, he commented. His research was often driven by chance: he would move from subject to subject, like a foraging bee, attracted by a publication, a discussion or, as in the next case, a letter from a physicist he did not even know.

Request for Reinforcements

At the end of 2005, Pierre-Gilles received a letter asking him to write the preface to a book on superlubricity. He was surprised, having never heard of it. Stung by curiosity, he discovered a remarkable phenomenon: when two graphite crystals are rotated together, the friction between them is abnormally low. Indeed, if they are held above a certain angle to each other, the so-called magic angle, the top surface layers of atoms fit together without deformation, and the two crystals slide over each other almost without friction. After further investigation, Pierre-Gilles realised that the theorists had provided explanations, but no overall picture. He was convinced that dislocations (discontinuities in the alignment of the atoms) in the graphite play a major role in the phenomenon and wanted to get hold of a book on the subject, *"the dislocations bible"*, as he called it, written by Jacques Friedel in the 1950s, which proved impossible to find. So he called the author to borrow his personal copy. The next day, he went to fetch the book from Friedel's home, and the two men spent part of the afternoon discussing the problem. Pierre-Gilles went off with the precious tome — it was a first edition, with numerous annotations by its author. He was careful not to damage, or worse, lose it, *"which would be an irreparable loss to history."* A few days later, he returned it to its owner. The two physicists considered the problem again, met several times, and finally demonstrated the existence of a low-coupling regime, which results in minimal friction (where they showed the importance of the surface waves), and a high spin regime, where "orthogonal screw dislocations" common to both crystals should condense at the contact plane, as the physicist Charles Frank had shown long before. *"We had a great time"*, smiled Pierre-Gilles de Gennes in recollection. The two physicists had not written an article together since 1969.

Sniffer Bacteria

Pierre-Gilles was unafraid to tackle any new subject. *"At my age, I'm not scared of any physics problem, whatever the field: hydrodynamics, quantum mechanics, statistical physics... I have enough experience to have encountered many different subjects"*, he asserted. As he got older, therefore, he mastered more and more fields and increased his chances of spotting unexpected analogies. The most striking example was bacterial chemotaxis, the ability of bacteria to move towards a food source (sugars, amino acids). Indeed, some bacteria, like *E. coli*, are sensitive to a chemical gradient and respond to it by changing direction. Pierre-Gilles showed that, in certain cases, they are attracted by aspartame rich regions, and remain in them, because it is something they produce themselves. The behaviour of these bacteria reminded him of... polarons, which he had studied 40 years earlier, prompted by articles by Richard Feynman. Polarons are electrons that cause deformation in the atom lattice around them and are then trapped, like the bacteria, by the deformation they create (the polarisation of the lattice acts like a potential well which prevents them moving). A novice would not have been able to see this parallel. It needed someone like Pierre-Gilles, able to span subjects as different as bacteria and dielectric crystals, to be able to do it. Nevertheless, he described the discovery of this astonishing analogy as a *"physicist's entertainment"*.

While recognising that he had become more effective with age, he also complained of a loss of concentration. *"Being old doesn't help: I'm getting more and more like Professor Cosinus. I make stupid mistakes"*, he grumbled. For example, he had begun calculations relating to the study of how small rubber balls slide and break away on a solid surface, following a remark by Françoise on the abnormally high force required to make these small balls slide on a table (compared with the predictions of the generally recognised model, called the JKR or Johnson-Kendall-Roberts model). She had also noted that the force employed to make them break away from the table depended on the speed of action: the harder one pulled, the faster they broke away — *"it's typical of Françoise to spot where a problem lies"*, he remarked. On reflection, he discovered that the adhesion energy depends not on the area of adhesive contact, but on the diameter of the ball, and he was on the point of sending his article to various specialists, when he was seized by doubt. *"I wondered whether I hadn't confused two 0s."* Mistakes are inevitable in any research. Like everyone else, regardless of age, Pierre-Gilles had made mistakes here and there. Unlike others, however, he was happy to talk about them, even going so far as to give lectures on their benefits in order to come down off the pedestal where people placed him. He was keen to stress that he too was groping in the dark, could make mistakes, start again, to encourage other people, in particular youngsters, not to be afraid of taking a wrong turn. *"Hearing things like that sets you free"*, comments Élie Raphaël, one of the physicist's former students.

When Solids Flow

In 2005, the whole solid-state physics community was in a fever about supersolids. Pierre-Gilles was not exempt, and began a long investigation of the subject after reading an article by his colleague, Philip Anderson, *"the pope of solid-state physics"*, published in *Science* in November 2005. Since the late 1960s, physicists had thought that the property of superfluidity (frictionless flow) is not confined to liquids, but can occur in solids (through frictionless interpenetration at very low temperatures). Indeed, in a crystal, displaced atoms are replaced by "gaps", which are mobile and change places with their neighbours: they are said to be delocalised. This delocalisation could give rise to a general frictionless movement that would make the crystal superfluid, or rather, in this case, supersolid. Some objected, however, that the gaps would inevitably vanish once they reached the surface of the crystal, which would make the phenomenon disappear as soon as it occurred (if it actually existed). *"As a result, interest in supersolids lapsed"*, explained Pierre-Gilles, who had himself studied certain aspects of superfluidity in 1973 (notably, the effects of anisotropy). However, new experiments undertaken at Penn State University in the US revived the subject in 2004. They showed that the moment of inertia (resistance to starting to spin) of solid helium is abnormally low, which characterises the appearance of superfluidity in liquid helium 4 when it is rotated. Could this be experimental proof of supersolidity? Pierre-Gilles doubted it, but was persuaded by a second experiment. Every solid-state physicist was on a war footing, Pierre-Gilles included.

His idea was that the phenomenon might not be caused by gap displacement, but instead by dislocation creep. Indeed, he suggested that the dislocations are mobile even at zero temperature, and that they jump from one site to another by quantum tunnelling. *"The dislocations do not move in a single block — the object is too enormous — but by 'detachment'"*, he explained, i.e. small portions of the dislocations combine. He wrote an article and sent it to a few colleagues, including Jacques Friedel — using the article that they had published together as a springboard. On further reflection, he became convinced that his idea did not explain the experimental results obtained by the Americans. So he changed tack and constructed a model in which the displacement of impurities on the dislocations through quantum tunnelling is responsible for the diminution in the moment of inertia, rather as unsecured packages in a train tend to accumulate on one side in bends. He sent this new version to several physicists, including Sébastien Balibar at the École Normale Supérieure. The two men made an appointment for lunch, delighted to work together for the first time in the 35 years they had known each other. *"In fact, I had made a mistake; I had read too quickly* (Author's note: in transcribing experimental data). *It was Balibar who pointed it out"*, recalled Pierre-Gilles de Gennes. He corrected his mistake. In the end, his thinking led him to a model which describes the dislocations in a quantum regime (without explaining the reduction in the moment of inertia of solid helium), which he sent to the *Proceedings of the Academy of Sciences*, in order to *"bank the idea permanently somewhere, without drawing attention to it or advertising."*

A Lonely Attic

His move to the Institut Curie had given him wings to move on to new subjects. It also brought him closer to people he liked (Jean-François Joanny's office was on the floor below his) and to Françoise. Nonetheless, he sometimes felt isolated. At the Collège, as soon as he had an idea, he would seek out the other theorists in the lab, Élie Raphaël or young researchers like Achod Aradian or Yoav Tsori, to share his thoughts and do some of the calculations. During this time, he would talk to the experimentalists or study other subjects. There were constant interchanges with his collaborators. At the Institut Curie, it was not the same. He would usually have to wrestle with his calculations on his own, and sometimes felt that he was tossing ideas in the air without seeing them land on a lab bench for experimental testing. *"As there is no experimental team nearby, and the experimentalists I know are working on different subjects, it doesn't lead to any activity"*, he regretted. He would spend more time in his office. This new office was as tiny as the one at the Collège de France was huge, but it was not without its charm, situated under the roof, with a big window opening onto the tops of the trees in the garden. As always, the theorist had arranged a pleasant workplace, surrounded by photos of people like the British physicist Charles

Franck, Hubert Gié and of his mother, dignified and upright in a dark coat. He corresponded regularly with Sam Edwards, who also complained of isolation, going so far as to say: *"Fortunately de Gennes is there."*

Of course, he still received regular visits, for example from former collaborators seeking his advice. Sometimes, in the course of a discussion, he would get up, saying: *"That reminds me of something."* He would rub his chin in front of the dozens of orange files covering an entire wall of his office, pull one out, and leaf through it for a few moments before picking out an article and handing it smilingly to his guest. Bingo! He would have found, amongst all the publications he had been keeping since the 1950s, the exact one relevant to the discussion — evidence, if needed, of his phenomenal memory.

His timetable was still very full. However, he turned down more and more invitations, withdrew from numerous company boards and scientific councils, and entirely ceased to set foot in places where he was already a rare visitor, such as the Academy of Sciences. Although he had reduced his commitments, he had to pay the price of his status and could not avoid all his obligations. It was, for example, in his role as president of the Centenary Committee of the Nobel Prize for Physics won by Pierre and Marie Curie that he welcomed Bernadette Chirac, Princess Victoria of Sweden and figures from the scientific and medical world to the great amphitheatre of the National Museum of Natural History in December 2003. But he did not restrict his attendance to the most prestigious ceremonies. On the contrary, he agreed to take part in events supporting causes close to his heart. In 2005, for example, he attended the general assembly of *Thésame*, the mecatronics innovation centre in Annecy, and in November 2006, the conference of the association of post-doctoral and postgraduate students of the universities of Aix-Marseille, to support their information processes.

The Characters of la Bruyère

Time did not seem to diminish his enthusiasm for science or for the ideas he believed in. He tirelessly continued to put across these messages, even if it meant going against popular opinion. *"Israel is a typical case. I find my position diametrically opposed to that of some of my colleagues* (Authors note: call for a boycott of Israeli research launched in 2003), *without claiming that I'm always right"*, he explained. He rejoiced in any progress *"in the right direction"*, for example at the introduction of autonomous learning practices in lycées (by the Minister Claude Allègre), with the aim of making education more pragmatic. In 2006, there was also a plan to attach two of the labs at ESPCI to INSERM instead of the CNRS. *"I have the impression that I struck a spark by introducing biology to ESPCI and that I am now seeing a vigorous flame."* It also made him very happy whenever young people thanked him for one of his lectures which, they claimed, *"had changed [their] life"*. One day, as he was arriving at the Institut Curie, a man hailed him in the street: *"Monsieur de Gennes, my son listened to your lecture three months ago and now wants to go into physics!"* When he realised that his efforts could occasionally bear fruit, he would say that *"it makes it all worthwhile."*

If he sometimes experienced disappointments, they did not make him bitter. So instead of getting irritated about the bigmouths and gasbags he had encountered in his career, he lampooned them in a series of short literary portraits, resembling la Bruyère's *Caractères*, in a book called *Petit point* published in November 2002. *"The book is sometimes harsh, but less harsh than my reaction in the heat of the moment. With hindsight, I consider the context and judge them more kindly, an attitude I have acquired with age"*, he confessed. These portraits are indeed pitiless — only two of them were favourable (those of Vinay Ambegaokar, who he had spent time with at Cornell, and of Lisbeth Ter Minassian Saraga, a physical chemistry researcher he admired) — but never vicious. The big game in the physics community when the book came out was to guess who was hidden behind the punning pseudonyms that the author had thought up. For example, "Breton" was inspired by Yves le Corre, a physicist who was president of Paris 7 University from 1976 to 1981 (though many others identified themselves as the subject). Pierre-Gilles claimed to have mischievously slipped his own portrait into the book, and found it funny when he was asked whether he was "Leduc" or "Robert".

In Saint-Tropez

Since he was now free of many obligations, he had more time to share with his two families. He would generally arrive in Orsay on a Friday, spend part of Saturday there, and then join Françoise in Saint-Cloud (or perhaps the other way round). Monday evenings would usually be spent with Annie in their smart little apartment on Rue Mouffetard. On Tuesday evenings, he would go to Saint-Cloud, but return on Wednesday morning to Annie in Rue Mouffetard. From here he could go on foot to the Institut Curie, taking advantage of the walk to think or entertain himself by estimating orders of magnitude, his *"morning jog"*, as he called it. He would spend Wednesday evening in Orsay, Thursday in Saint-Cloud, and so on. *"In the mountains, I have a base at Orcières with Annie and another at Serre-Barbin with Françoise, a house we have been renting from friends for 30 years now"*, he reported.

With Annie, he was in clover. She still took care of everything and spoiled him. *"You do everything so well. Why would you want me to get involved?"*, he would say. She had arranged a spacious den on the first floor of their house in Orsay, a place to relax, where Pierre-Gilles would read, sometimes science-fiction, provided that it was not *"too scientific"* — *"What we like is the appeal to the imagination and a certain vision of the way societies work, ideas about the disruptions, the collapses that threaten them"*,[2] he explained — listen to music, draw or paint. Indeed, after having long been satisfied with line drawings, he had now ventured into paint and even had lessons every week. *"My mother laid on drawing lessons for me, where I learned perspective, one of the basic skills: you have to learn it, understand it and then forget it. It's very useful to me: I save masses of time compared with the other students. Some of them find the problems of perspective very hard, but not me, whether I'm drawing a hilltop village in the Midi or a woman — my two favourite subjects."* But while he usually considered his sketches fairly successful, he was rarely satisfied when he moved on to the painting stage. *"I'm too clumsy for painting with a palette knife. And I don't always achieve*

---

[2] *Elle*, March 16, 1992.

colour harmony. From a distance, it's OK, but not close up", he mourned. His models were Botticelli and Nicolas de Staël for colour, Dürer for drawing.

Annie and he also spent some fine moments in the South of France, for example in 2005, when they organised a big picnic with all their children and grandchildren on a plot in Saint-Tropez that Pierre-Gilles' father had bought in 1928 and that they had never sold. *"It's located above Salins beach and the Bay of Pampelonne. It's a magnificent spot"*, reports Annie. The whole family ate a cheerful lunch, in the shade of the pines, before starting a long walk on the smugglers' path to the beach, where the younger ones went swimming.

In his other family, he was (slightly) more involved in day-to-day tasks. For example, he would lay the table with great care, favouring an asymmetric composition or adding a flower. He spent time with Marc, the *"baby of the family"* — the older ones had now left home — sometimes taking him to the swimming pool. On Sunday mornings, he and Françoise would usually go to an exhibition, such as Pierre Bonnard's show at the Paris Museum of Modern Art in 2006. They would browse until early afternoon, then have lunch. After this, they would jot down their impressions in notebooks. When Mathieu, now preparing a doctorate, was at home, there would be endless discussions on theoretical physics — *"what bliss!"*, smiled Pierre-Gilles de Gennes — which would drive the other members of the family away.

A Long Illness

"The fact is that I have had a long series of health problems — the most recent being hepatitis, which left me little time to work during the day", he wrote to the director of the Collège de France in 2001. *"As a result, I am not sure that I will be able to prepare a course for next year. If things improve, I will do everything to provide one, but if my current condition persists, I will have to cry off."*

When he was diagnosed with cancer, Pierre-Gilles initially suspended his plans. For example, in 2002 he declined the offer by Vance Bergeron, a former researcher at *Rhône-Poulenc*, to act as a scientific adviser for his new company, *Air In Space*. He explained that, being about to start chemotherapy, it would be *"unwise"* to commit himself, not knowing whether he would *"still be in a condition to perform such a role three months later"*. However, he quickly picked up the threads of his life, trying to reveal nothing of his condition. *"His lecture was, as always, amusing, aesthetic and dynamic. He showed no trace of the cancer he was rumoured to be suffering from. And, although he seemed tired after his paper, he was as considerate as ever in answering the endless questions that followed"*, reported a female physicist in 2006.[3] He had highs and lows. Usually equable, his mood would sometimes darken, becoming *"a little gloomy"*, says one female friend.

The disease was accompanied by the asthma attacks, which he himself described as *"terrifying"*, which had haunted him for years. The worst had taken place a few years earlier, on April 20, 1995. He had had an exhausting day, delivering the inaugural speech in the ceremony marking the transfer of Pierre and Marie Curie's ashes to the Panthéon. *"An icy wind was blowing"*, he recalled, while the students slowly advanced up Rue Soufflot, carrying the two coffins to the esplanade decorated with a gigantic French flag. Pierre-Gilles was frozen when he began his tribute, broadcast by loudspeakers as far as the Luxembourg Gardens. He paid tribute to Pierre *"both observer and thinker, but at the same time a craftsman (...), who built apparatus of which every example is a model of refinement"*, then Marie, *"patient, reflective, persistent"*, who had given the name *"of her oppressed and beloved country to the first element"*, he had said, turning to Lech Walesa, present on the official platform, alongside the family, François Mitterrand and Édouard Balladur. He had finished by speaking of *"Marie's last great pleasure"*, the discovery of artificial radioactivity, and concluded his speech on our societies' responsibilities on nuclear energy: *"What comes next is our problem."* The ceremony had exhausted him. That same evening, he caught an overnight train to Briançon to join Françoise in Serre-Barbin, in the Hautes-Alpes. Around midnight, however, he experienced an asthma attack that his medication was powerless to control. He managed to get out of his compartment, drag himself to the inspector's

[3] Anita Mehta, *The Hindu*, May 31, 2007.

cabin and call for help by banging on the door. The inspector had immediately raised the alarm, and when the train arrived at Valence, an ambulance was waiting for him. *"It was one of the greatest panics of my life. I thought the end had come... It's lucky it happened before Valence"*, he remarked.

In May 2005, he visited his old friend Jacques Labeyrie, *"an extraordinary storyteller"*, in his farm near Fayence in the Var region. With his children Christian, Dominique and Marie-Christine, they were enjoying the stories told by this former companion of Haroun Tazieff, when he suddenly felt ill. It was an allergy: he hadn't noticed that there was a poodle in the house. He spent the remainder of the afternoon in his room, trying to catch his breath.

Despite his asthma attacks, he continued to smoke. He had tried giving up just once, following a health problem in 1997, at the demand of his daughter Claire. Driving to Reims to give a lecture, he was forced to turn back by terrible pains. In the ER, he was operated on for appendicitis. But the pain had continued: it was in fact a kidney stone. He was relieved, but he had a scare. His family too. Claire nagged him to give up smoking. He accepted what he jokingly called his daughter's *"vicious blackmail"*, and stopped smoking for several months, but then took up the habit again. He was never without his cigarillos. On one occasion, when he was giving a class at the Collège de France, a physicist had pointed out that smoking was forbidden. Furious, he snapped at her: *"Anyone who doesn't want to follow the course can leave."* When, 20 years later, a colleague reminded him of the incident, he replied that he had made it a matter of principle, as he explained on the programme *L'Heure de Vérité* in 1992: *"I know that smoking is dangerous. I don't advise anyone to do it. But I am also very shocked by this society with its polarised ideas of right and wrong."*

At the beginning of 2005, he took part in the launch of Physics Year at Lyon City Hall, then left to return to Paris. One of his physicist friends walked him out and asked about his health, worried to see him so frail. Pierre-Gilles waved the question away. His friend pressed him: *"But you're having treatment?"* *"Yes"*, he smiled. *"I hope you're going to stop smoking."* Pierre-Gilles chuckled and said: *"Ah, you reminded me"*, taking a cigarillo out of the packet.

Concerns

Periods of remission alternated with spells of pain. In 2006, in the USA with Annie to attend a prize award for his colleagues Bob Meyer and Ludwik Leibler, he was awoken in the night by pain. He wanted to leave the US immediately, *"because of the extra examinations that hospitalisation there would entail"*. He flew back to France the next day, stuffed with painkillers for the flight. Fortunately, the pains were only the symptoms of a bad case of flu.

As a well-known scientist at the Institut Curie, he received letters from desperate patients. *"At the moment, I am in chemo myself, and I fully understand your anxiety. But at Curie, I am only a lowly researcher: I have no knowledge or information about the treatment of recurrence. I believe we need to trust our own doctors. With all my sympathy..."*, he wrote in December 2004. He trusted his doctors, but wanted to avoid *"over-medicalisation"*. In 1999, even before he became ill, he had joined the ADMD (Association for the right to die in dignity), signing a decla-ration of civil disobedience: *"We declare that we have helped a person to die or would be ready to do so. We believe that the freedom to choose the time of one's death is an inalienable individual right, inherent in the Declaration of Human Rights."* He went on to explain his position: *"I don't feel in a position to judge, because I have had neither the energy nor the time to study the question in detail. However, my interest is essentially personal, because I would very much like not to receive excessive treatment at the end of my life, nor to be kept alive artificially for no purpose. And with the type of condition I have, the end can be very painful. So I would like the law to be much more liberal and allow you to talk with your doctor about ways of "making things easy."*

It was a time when his views on religion changed. He had never been a churchgoer, but he recognised the role of protestantism in his upbringing, for example his time with the Morch family in the late 1940s and early 1950s. He had his children baptised and regu-larly read them the Bible. But so far, he had tried not to mix science and religion *"I think it is clear that science and religion should be seen as two separate mental domains. I tend to believe that our mind has needs on both sides — rational and irrational."* He explained this quest for spirituality — for the irrational — by citing the theologian France Quéré, wife and mother of the physicists Yves and David Quéré: *"Our brains work like electric railways. We think the same way we travel by train, on the same tracks and stopping at the same stations. These beaten tracks distance us from the mysterious travellers that we really are deep down — having been born free."* But in 2005, in response to a journalist's question, he replied: *"Today, my answer is more uncertain. I believe that in the course of this century, research on the functions of the brain will tell us whether human beings have free will or not. If it turns out that we have not, religions will cease to be relevant. However, it would be premature to try to prejudge the issue. Especially as history has shown that the judgement of both scientists and philosophers has often been completely wrong. You only have to think of Kant or Sartre."*[4]

Farewells

He went to see many of those who had been important in his life, as if to say farewell to the people he cared about. There are too many of them to list, but, for example, he made the effort to see the solid-state physicist Charles Kittel, now aged almost 90, who was tending his vines and producing his own wine in Northern California. He got the opportunity to revisit the campus at Berkeley *"with some emotion"*.

In May 2006, Phil Pincus organise a symposium in his honour in Trieste, bringing together a small group of some forty scientists. It was not the first conference dedicated to him. The biggest had marked his retirement from the Collège de France in June 2002, which was attended by an impressive number of colleagues. However, the Trieste event was special. Pierre-Gilles was surrounded by his closest collaborators, who all knew he was sick. He had given his agreement for the meeting, *"provided that we don't go over the past"*, but that discussions were confined to the most recent scientific results.

---

[4] Bernard-Olivier Schneider, *Le Nouvelliste*, August 25, 2005.

Despite his illness, he did not take things easy. In fact, he seemed to work harder at times of difficulty. In December 2006, he was *"tense"*, because the news about his health was *"not reassuring"*. All the same, he spent Christmas in New York with Françoise and their children, then celebrated the New Year with what he called his *"big children"*. And he quickly returned to his research on quantum regime dislocations. *"I'm very lucky, because this type of new object of curiosity means that I don't have too much time to think about my own fate; it's excellent"*, he commented. He was also beginning to take an interest in earthquakes and was closely following Étienne Guyon's studies on honey flows. In 2007, however, he would be entirely captivated by a new subject.

Final Publication

"I have just had a weird idea", he wrote in January 2007 to his friend Guy Deutscher. After explaining it, he asked him what he thought and *"whether anyone else had already come up with it"*. Guy Deutscher was amazed: Pierre-Gilles had just thought of a new mechanism that would explain nothing more nor less than high-temperature superconductivity. *"I'm asking your advice, because I don't have the necessary background"*, he mentioned in the message to his colleague, who had been researching superconductors since the time of the "four musketeers" at Orsay.

This was an unexpected U-turn from Pierre-Gilles de Gennes. He had lost interest in the question back in 1987, thinking that there were no new physical effects in high-temperature superconductivity, but that it essentially arose from a combination of several well-known phenomena. In private, he went so far as to say that researchers were wasting their time on the subject. Nonetheless, he had always followed what was going on in the field. *"There are thousands of ideas, but none of them is convincing. It's a can of worms"*, was his view. And here he was, suggesting that there was a new effect in the family of high-temperature superconducting oxides. His idea was that electron pairs are responsible for superconductivity in copper oxide, just as they are in classical superconductors (which is contested by numerous physicists), but their formation is triggered by the overlap of the atomic orbitals, i.e. the electron clouds orbiting the atomic nucleus (and not by lattice deformation as in the case of classical superconductors).

Guy Deutscher's initial reaction was scepticism. One step in the process seemed unconducive to the superconducting transition. He shared his observation with Pierre-Gilles, who amended his description. The two physicists worked together to refine the model, then submitted an article to the *Proceedings of the Academy of Sciences* at the beginning of April. A month later, they got the feedback from the peer reviews. The first was positive, but the second incendiary, with the reviewer claiming that it was not a new idea and citing references. The two physicists agreed to meet again at the end of May, to draft a response.

But Guy Deutscher would have to correct the article alone. *"This idea was so important to Pierre-Gilles that I had to continue"*, he explained. He included the references in question, explaining why they were not relevant to this particular case. The more he reflected, the more convinced he was that the mechanism could be correct. Indeed, it also explained why currents do not travel well between grains[5] in high-temperature superconductors, as they do in classical superconductors, a difference that was still not understood. With the model proposed by Pierre-Gilles, it is all self-explanatory: the orbitals do not overlap so well, because of the grain boundaries, which undermine the formation of the electron pairs. In addition, in classical superconductors, a certain characteristic length called the "coherence length" is much greater than the inter-atomic distances, which favours collective phenomena. On the other hand, in high-temperature superconductors, the order of magnitude is the same, which confirms the significance of the local structure and in particular the grain boundaries. This new model also explains why the critical temperature in the copper oxide family varies with the presence of other ions. The introduction of these ions changes the

---

[5] Grains are minute crystals that constitute polycrystalline oxides. They are separated by so-called grain boundaries.

angles between the atomic planes, hence the overlap of the orbitals. And it turns out that the critical temperature is indeed at its highest in the compound where the overlap is greatest. Pierre-Gilles de Gennes' final article would eventually be published in November 2007. It would go relatively unnoticed. *"A lot of work needs to be done to confirm the mechanism. But no prohibitive argument has so far been put forward"*, notes Guy Deutscher.

The Final Months

The final months were very hard for Pierre-Gilles de Gennes. The most powerful painkillers no longer brought relief. Usually a very early riser, he would get to his office later, and sometimes leave in the afternoon to rest. He did not try to hide his condition, but nor did he speak about it, just referring to it when necessary, saying when making an appointment: *"if all goes well"; "God willing"* — or more jokingly — *"Careful, I'm radioactive this morning"*, as he said smilingly, after a scan. *"I try not to be boringly afraid of death"*, he would say.

Chapter 16

France's Pays Tribute

In late afternoon of Wednesday, May 16, 2007, as was his custom, Pierre-Gilles de Gennes tidied his office before leaving. It may be that he had been a bit more careful in recent weeks, and that he had made more effort to make sure that his bibliography was up-to-date. Again as usual, he had spent the day working. He had also confirmed the starting time of shooting for a documentary on his career, due to begin the following Monday, made appointments for June and given his agreement to join the Scientific Committee for the centenary of the newsletter of the Union of Physics and Chemistry Teachers in October 2007. On the other hand, he had withdrawn from the retirement ceremony for the physicist Vinay Ambegaokar, in mid-June in the USA, pleading tiredness. In short, it was a day like any other. When his secretary Marie-Françoise Lancastre left, they said *"see you Monday"*, since it was the beginning of the Ascension holiday. Pierre-Gilles stayed on for a while, then went home to Orsay.

On the Friday morning, Annie was getting ready to leave for a funeral in Montargis. She did the chores, got dressed and left. Pierre-Gilles sat down at his desk to work. He died that morning.

Rich Legacy

Pierre-Gilles de Gennes was one of the the 20th century's most significant physicists, not only because he revitalised physics by opening up new fields, but because he approached it "impressionistically", by means of scaling laws. He left books that remain seminal works. On the other hand, he ultimately trained few students, responding to the numerous requests to supervise PhDs or postdoctoral work: *"Unfortunately, we are full up next year"* (though he would then recommend other research groups). Nevertheless, he left his mark on many physicists who pursued similar research or adopted his style.

As the author of more than 550 publications, he left a significant scientific legacy, which remains to be fully assessed. Some of his predictions and models still await verification. For example, the classical Casimir effect that he predicted in 1978 with the physicist Michael Fisher was only demonstrated directly in January 2008. He believed that his most important discoveries were: the HC3 field in superconductivity; the idea of reptation and the n = 0 theorem for polymers; the triple line dynamic in wetting. Of all of the subjects he tackled, there were some for which he had a particular fondness, such as turbulence, whose mysteries he dreamed of resolving in 1966, then in 1986 with David Tabor, of the UK's Cambridge University, without ever doing so. Chirality was another of his favourite subjects. Chiral molecules are asymmetric molecules which, like people, can be right- or left-handed, and can modify the properties of a compound in which they occur.[1] Employing an argument from quantum physics, Pierre-Gilles had demonstrated that, contrary to Pierre Curie's 1894 finding, an electromagnetic field does not promote the occurrence of right-handed or left-handed molecules. He would return to this subject several times in his career, in different contexts (bilayers, crystalline structures, etc.).

---

[1] For example, a nematic liquid crystal composed of chiral molecules becomes cholesteric and has different properties.

A career like the one Pierre-Gilles enjoyed seems impossible today. Firstly, today's researchers are more specialised. Indeed, there are fewer fields to investigate in physics, and they are now patrolled by armies of researchers, whereas for Pierre-Gilles and his generation, there was so much to explore! The theorist of soft matter belonged to that generation of physicists whose culture embraced all, or almost all, of physics: their bible was the six volume treatise on physics written by Lev Landau and Evgeny Lifchitz (which covered both mechanics and quantum electrodynamics) and they could find everything that had been published in a journal like *Phys. Rev.* Moreover, physicists of Pierre-Gilles' generation could work without worrying about the future. The autonomy they experienced is no longer available. These days, young researchers *"work under pressure"*, needing to publish to get recognition, needing to focus on *"fashionable"* subjects and to find funding, a trend that Pierre-Gilles de Gennes regretted.

Decipherment

Pierre-Gilles de Gennes rose to the highest level through immense intelligence and an extraordinary memory, but a physicist's success does not depend solely on his intellectual capacities. Other qualities are needed, which distinguish an excellent researcher from an exceptional scientist like Pierre-Gilles de Gennes.

The physicist of soft matter was above all driven by an inner fire that could only be assuaged by the pleasure of discovery. *"Initially, I am fascinated by what I find and, for a while, all that interests me is the result. Then, I experience a second form of satisfaction when I turn to the experimentalists, and they devise new experiments. It's the satisfaction of... seeing! And then, after that, there is the pleasure of going to conferences and telling the story and realising that people are interested and it's something they had never thought of"*, he explained. It is hard to appreciate the sense of exaltation that comes from a discovery. *"What is sad about our sciences is the difficulty of conveying the feeling. Three pencil strokes by Picasso, a musical phrase by Vinteuil, can move people, but it takes long years to feel the beauty of a new idea in physics; the dazzling experience of a change of perspective which suddenly changes our understanding of part of Nature"*, he wrote in 1990.[2]

He also had a buccaneering spirit, which first prompted him to to move away from the beaten track of solid-state physics and then impelled him to move on again and again. *"[This thematic mobility is] crucial if you don't want to die ignorant, or at least complacent (which is perhaps worse). However, it is not the way to build a reputation. You have to be really good to be able to combine the two"*, commented Suzanne Quiers, a friend, in a letter she wrote to him in 1989. *"Moving on from a subject prevents you getting caught in a rut or becoming a mandarin, lording it from his ivory tower"*, is how he explained his approach, though he too experienced occasional doubts. *"To have many curiosities is nice; but to have several specialties is like having none."* This epigram, coined by the historian Maurice Agulhon on his departure from the Collège de France, had struck home with Pierre-Gilles, who wrote to him in 1997: *"We hardly know each other. But I recently read (in a hospital bed) your final lesson (...). I have constantly come up against this problem, in the course of a long and awkward journey through physics. At least I have recognised my mistakes (which were many). And I have tried not to be dogmatic."*

This refusal to settle into the comfort of a single subject was a reflection of his stringent self-discipline. Perhaps it came from his upbringing, which had left no room for self-indulgence and had always driven him to be the best? This self-discipline was embodied in his colossal capacity for work. He worked furiously at the beginning of his scientific career, aware that the early years were critical. *"Though I did know how to enjoy myself"*, he countered. After taking charge at ESPCI, he worked very hard again. *"I didn't have enough time to think during the week, because I was actually quite busy with meetings and other business for the school. So I worked at weekends"*, he explained, before correcting himself: *"I would still occasionally stop work to draw for an hour or two. It is good for the brain."*

---

[2] Martine Poulain, *La bibliothèque imaginaire du Collège de France*, Le Monde éditions, 1990.

He was also a methodical man, both in day-to-day life — where he was organised and extremely punctual — and in his thinking, with a talent for simplifying things. This was evidenced by his style. The style of his lectures — clear and uncluttered, like the transparencies instead of trendy, colourful presentations — of his articles, his letters, even his drawings. *"It is sketching I like the most. To render meaning, expression, in a few pencil strokes, makes you feel fantastic"*, he confessed. He always sought to reveal the essence, to strip away the superfluous, even to do DIY physics.

Unlike with some, his capacity for abstraction did not alienate him from the practical. Firstly, he kept his feet on the ground, a long way from the popular cliché of the absent-minded professor. He was also able to grasp the opportunities afforded by new experimental techniques developed in the course of his career (neutron scattering, tunnel junctions, etc.). He knew the practical details of certain experimental setups almost as well as the people who built them. He would spend ages examining the results of an experiment, as intrigued as if it were a five-legged green elephant. These are the characteristics that enabled him to make maximum use of his work with experimentalists.

Many physicists stress the intuition he showed in picking out subjects or finding solutions. Those lucky enough to see him reasoning aloud knew that his thinking could sometimes be hard to track, taking unexpected and tortuous paths. His thoughts might seem to have no connection with the initial question. But out of this apparent confusion would emerge an idea whose relevance might become clear to his interlocutor only a few weeks later. But does intuition truly play a role in a scientist's mind or was this quality supposedly possessed by Pierre-Gilles de Gennes not instead indicative of his immense treasure house of knowledge about physics?

To be successful, though, it is not enough to develop theories, however elegant. People have to hear about them. Pierre-Gilles de Gennes was a consummate publicist, travelling to every corner of the US, sending out dozens of preprints and diligently cultivating his network of contacts. His charisma and skills as a communicator did the rest. It was a talent not given to everyone, for example his friend Shlomo Alexander. *"He was a very deep thinker, but in his presentations he would reconstruct his entire reasoning processes on the board, without notes: he was impossible to follow!"*, smiled Pierre-Gilles de Gennes. Pierre-Gilles was someone people talked about, were attracted to, to the point that he attained an ascendancy of which he was completely unaware, perceived as an oracle despite himself. People wanted to be close to him, to be noticed, be acknowledged by him, were proud to have held a discussion with him. It bugged him, but what could he do about it?

The Failings of Genius

Not everyone felt this way about Pierre-Gilles. *"To my knowledge, he had no enemies, only people who were jealous, or very formal theorists who didn't like what he did"*, affirms one of his students. Nonetheless, he could antagonise people with sweeping statements, despite his insistence on providing evidence for his opinions. He could state harsh truths with disconcerting bluntness, which turned more than one colleague against him. However, it should be stressed that he showed the same spontaneity in expressing positive views. *"He would listen to a presentation, learn something new and declare the lecturer a genius, when in fact everybody else had heard it before"*, some claimed. He could also be disarmingly honest, even about his own articles. *"Overall, this article has two opposing deficiencies: it is completely lacking in mathematical rigour and, conversely, it has nothing to do with the practical sides of polymers, but there is a tiny possibility that it could help bring these two positions together"*, he remarked in conclusion to one of his articles on percolation.

On closer examination, it seems likely that Pierre-Gilles was not entirely at ease with people, was perhaps even shy. *"At ESPCI, people accused him of being remote. While he seemed warm on TV, he was perceived quite differently at the school. They thought it was arrogance, I think it was shyness"*, concludes one colleague. More out of modesty than any shyness, he was reluctant to talk about himself and tended to hide behind preprepared speeches or stock anecdotes. He could also be quite domineering when uncomfortable, and be unsociable to the point of leaving a reception or a family gathering without apology.

In fact, he was unapologetically himself, whether in work or in life. In work, he made no attempt to keep his models for his own experimental teams, prompting some to say that he did not really have a group, or else that the whole world was his group. (In his defence, it should be said that he changed subject on average every three years, whereas it takes an experimentalist at least twice as long to change course. So it was an impossible match). He would abandon a subject without worrying about the future of the people he left behind, though he would say that they were old enough to stand on their own feet. *"It's like with children, the best way for them grow up is to tiptoe away"*, he believed. In life, work always took priority over the wishes of those close to him. His family life suffered from sacrificed Sundays and incessant travel. While he had hundreds of friends around the world (some would say, "no one dared to be his enemy"), he only had two close friends, Phil Pincus and Shlomo Alexander. People around him would sometimes feel that they were interrupting something more important. *"Friendship takes time and he was never prepared to waste it"*, interprets Annie.

His individualism was so intense that he sometimes seemed to set his own rules. On the one hand, this made for a very unusual personality: a freethinker who, in his work, was not hindered by boundaries, whether between disciplines (physics, chemistry, biology...), or between public and industrial research, and did not give a fig for fashions in science or the prestige of scientific journals. *"[Bibliometrics] is not an exact science. (…) In France, we have Pierre-Gilles de Gennes, who scores poorly in this respect, though his work was of the greatest significance"*, commented Jean-François Bach, permanent secretary of the Academy of Sciences.[3] In life, he ignored convention and allowed no one to dictate the way he thought. On the other hand, he could sometimes give the impression of being above the law. He would smoke in lecture rooms. He chose openly to maintain

---

[3] Alain Perez, "In the jungle of scientific publications", *Les Échos*, May 22, 2007.

two families, causing suffering to those who cared for him. How much was he aware of this? It is a good question. One evening, in 2006, he had said to Annie: "Ultimately, I've been faithful to you." Chutzpah or naivety?

Nonetheless, he inspired respect from everyone, even his detractors, with his tireless and contagious enthusiasm, his immense generosity (he did not keep his ideas to himself) and a certain humility. "If I hadn't found these effects, someone else would anyway have found them soon after", he would say. He never boasted: he did not need to, the facts spoke for themselves.

He enjoyed a life full of intensity, punctuated with rich encounters — his regret was not to have known either Landau or Feynman, the two physicists he admired most (he felt closer in personality to the imaginative and original Feynman than to Landau) — and with travel. He heard the poems of Pushkin recited in St Petersburg; explored the Sinai desert; scaled the glaciers of New Zealand; sketched in the temples of Japan; contemplated the Tuscan hills of Villa Bencista near Florence... His curiosity did not stop at the walls of his study, but extended out into the world beyond. Whenever he arrived somewhere new, he would go out to explore. He was incapable of idleness, as if he had bulimia of the brain. He was a consummate multitasker: he would watch television with one eye while playing scrabble, reading a magazine and listening to a radio programme. His life was divided between two families; his work between two jobs (at the Collège de France and at ESPCI). It was as if he lived like a child learning to ride a bicycle, pedalling a top speed to stay upright.

Official Ceremony

His death was made public on Tuesday, May 22, 2007. Official messages and tributes poured in — *"Embodiment of scientific excellence"; "Above all a free spirit"; "One of the greatest physicists of our age".* There were those who sought to emphasise their link with the dead man — *"I had the good fortune to work with Pierre-Gilles de Gennes when I was Minister of Health",* noted Bernard Kouchner, Minister of Foreign Affairs; *"I have not forgotten the support he gave me in the 2002 presidential campaign"*, stressed Jean-Pierre Chevènement, former Minister of Research.

The Academy of Sciences was thinking of organising a tribute. Claude Allègre put in a word with Nicolas Sarkozy, recently elected on May 6. The former minister was a confidant of the new President, whom he declared himself "ready to help" (having turned down a ministerial post). The idea of the tribute was approved, but speed was of the essence. The deadline was to organise a ceremony before the first round of the parliamentary elections on June 10, 2007, and the date was finally set for June 5. This would allow the new President to kill two birds with one stone, pay tribute to "France's favourite physicist" and address the research community before the elections. Officially, it was organised by the Academy of Sciences at the Palais de la découverte, but in fact it was orchestrated by the President's office.

The Ceremony

On Tuesday, June 5, the guests converged on the entrance to the Palais de la découverte, but were held back. The sniffer dogs of France's anti-terrorist unit, RAID, were checking every corner of the building. No one was allowed in until the security services had completed their inspection. Sitting at a nearby terrace, the Nobel prizewinner Claude Cohen-Tannoudji wore an expression of infinite sadness. The guests massed on the steps and waited in the sunshine. Amongst them were former academicians and members of the family, even they prohibited from entering. The queue lengthened, covering the whole pavement.

The ceremony venue was an excellent choice. Pierre-Gilles had always been very fond of the Palais de la découverte. He had set foot there for the first time in 1945 and was still going to exhibitions in 2004. His presence on that occasion had not gone unnoticed, since a few days later he received a letter asking for his impressions. *"Electrostatics: clear and simple presentation (...). The experiments are still just as good (our son Marc, aged 14, long hair, joined in!)"*, he had replied. He had pointed out an error, *"a detail"*, and concluded: *"Once again, bravo and thanks to your presenters, who possess both science and enthusiasm!"*

He had always supported the Palais. On November 25, 1993, he had written to François Fillon, Minister of Higher Education and Research, to *"register concern"* about the plan to move the institution. *"The Palais is an instrument of scientific education at least equal to its rival at la Villette. (...) It would be shocking if the sciences were to lose their educational home in the heart of the city"*, he had pleaded. Two years later, he had signed an article in *Le Monde* headed *"Save the Palais de la découverte"*, noting: *"Many of us discovered our passion for science there."*[4] The Palais de la découverte had even exhibited his drawings, which were hung alongside works by Pasteur and other scientists in an exhibition on art and science, prompting him to wonder *"whether it was not slightly impudent"* on his part, even though there were drawings there *"which he would not disavow"* (he had thrown away entire armfuls).

The ceremony was running late. When the security services finally finished, the guests were in a hurry to enter. They had to show their credentials — their official invitation — to go through the security gates and access the immense entrance hall. A blue backcloth, overhung by a large photo of a smiling Pierre-Gilles at a blackboard, dominated the platform. To the left of the lectern hung a European and a French flag. In the front row, opposite the platform, sat the children of the Nobel laureate. The assembly brought together the cream of France's scientists. The former researchers from the Collège de France lab, Élie Raphaël, Liliane Léger and the others, were also there, spontaneously gravitating to the same row, though as they were not part of the *persona grata* of French research, they had almost not been invited.

Catherine Bréchignac, President of the CNRS, officiated as master of ceremonies. She announced that on Nicolas Sarkozy's arrival, everyone was to stand up and the speeches would be interrupted to give the floor to the head of state. Nicolas Sarkozy was in the process of unveiling a plaque in the theorist's honour at ESPCI and renaming the area dedicated to popular science *Espace Pierre-Gilles de*

---

[4] *Le Monde*, February 9, 1995. The Palais de la découverte is more than ever under threat of closure.

Gennes, in company with Valérie Pécresse, Bertrand Delanoë, Claude Allègre and Annie de Gennes. The atmosphere in the hall of the Palais de la découverte was neither restrained nor attentive. People were chatting, looking round, anticipating the arrival of the new President.

Jacques Friedel began his speech. As a friend from the early days, he retraced Pierre-Gilles' career, concluding: *"In recent years, he was trying to solve what he called small problems (...). His mind was always lively and acute; I had the great pleasure of resuming our discussions on subjects of shared interest — the plasticity of different crystals, then superconductivity, a subject he was still phoning his student and friend Guy Deutscher about just a few days before his death. He was, I think, a great scientist."* Étienne Guyon succeeded him on the platform. *"All the subjects he touched became, as if by magic, topics of interest, contributing to research all over the world"*, he commented. *"He exercised an attraction on everyone around him"*, he added.

Annie de Gennes and Claude Allègre made a discreet entrance, while the President lingered at ESPCI. Jean-François Joanny began the third tribute. He expressed sympathy with the family and reminisced: *"I remember those moments of intense excitement when he was focused on a problem (...). He would run into our office to explain his new ideas, once, twice, three times a day, though we hadn't yet grasped the first idea. I had no inkling, at the time, what influence he would have on my life. (...) For many of us at the Institut Curie, he was something of the Great Sage; very discreetly, he would talk to all of us and suggest new and constructive ideas. I never saw him turn down a scientific discussion, especially with students, when they had the courage to talk to him."*

The speeches continued against a background of low, but continuous hubbub, punctuated by the buzzers of the security gates that sounded every time anyone entered. There was a degree of agitation in the auditorium. Nicolas Sarkozy had still not arrived.

Phil Pincus took his turn. He was very emotional. He described their very first meeting in 1959, then commented: *"Most scientists feel uncomfortable talking about their work with nonscientists. Pierre-Gilles was different. His wonderful ability to communicate, both in French and English, together with his scientific talent, catapulted him to international celebrity."*

Philippe Nozières then took the floor. He told how, a few years before, he had reminded Pierre-Gilles of the subject he had chosen at the end of the Summer School in Les Houches, entitled "Adiabatic processes in thermodynamics and quantum mechanics". *"He had burst out laughing, saying: "What on earth did I find to say on a subject like that?"* Philippe Nozières was then interrupted by Catherine Bréchignac, announcing the imminent arrival of the President. As per instructions, he made to leave the platform. She stopped him: *"No, stay there, we'll follow protocol."* Everyone was standing, but the President did not appear... False alarm. She invited Philippe Nozières to continue. The audience sat down. *"Pierre-Gilles was an explorer, almost an adventurer, more inclined to slash himself a path with a machete than to cultivate a country garden. He would have been at home in the great 18th-century era, when the Academy of Sciences sent a team of scientists to South America, supposedly to measure the meridian, but in fact to find out everything about the unexplored continent"*, he continued. He stopped again. This time, Nicolas Sarkozy had really arrived, surrounded by a mob of journalists, assistants and bodyguards. The President shook hands with people along the way, then with the family in the front row, and sat down. General embarrassment. Catherine Bréchignac intervened: *"Given the time available, Philippe Nozières will finish there, um, has finished..."* There were protests from the auditorium. *"Alright, he'll continue"*, prompted the President of the CNRS. Laughter and applause. Philippe Nozières picked up the thread of his speech. Then, she gave the floor to the *"President's guest"*, Claude Allègre. There was further disturbance in the hall as some thirty members of ESPCI made their entrance, delayed by the ceremony at the school. Then Nicolas Sarkozy began his address. He paid warm tribute to Pierre-Gilles de Gennes, summing up his career, before continuing: *"As President of the Republic, I feel a twofold duty to Pierre-Gilles de Gennes: that of keeping the promises made to revitalise our higher education and research system and elevate it to the foremost international standards; and that of bringing the immense legacy left by our most popular Nobel laureate to fruition."* Christine Albanel, Minister of Culture and Communication, was also present. *"Yes, I will give higher education and research the resources they currently lack. Yes, I will carry through the profound reforms we have so long failed to implement. (...) Yes, I will introduce a plan to restore working conditions and salaries in research, particularly for young scientists"*, he promised. And the role of the sciences in education? *"In the plan that I will soon be setting out for the future of education in France, I will give absolute priority to the awareness of science, of its spirit, to the sense of its grandeur and beauty. I want our schools to give our children the taste for observation and the curiosity to understand with which any scientific vocation begins. I want our schools to give our children back the joy of learning and the taste for knowledge as a reward after the long effort of thinking"*, he went on.[5] Finally, he announced that the campus of the Orsay faculty would bear *"the great name of Pierre-Gilles de Gennes"*.[6]

"I want to say that in my eyes, France has no more precious assets than its scientists, its researchers, its great scientific tradition, and that in today's world, I am convinced that it is science on which our future depends. I am here to tell you, the State will be at your side to help you, to support you, to give you the resources to seek and to find."

The audience was stunned. The political message was all too transparent. The President could have put aside his soapbox for the day. However, it was too good an opportunity, in front of this assembly of the greatest names in French research, to announce his electoral promises! All the press reports on the ceremony would focus on nothing else: only his assurances about *"the profound reforms"* that he would introduce and his *"high ambition for French science"* were reported. The political message was heard loud and clear. The tribute, tarnished by this clumsy attempt at political exploitation and by the kind of fairground atmosphere that Pierre-Gilles de Gennes hated, left many of his friends and family with mixed feelings.

---

[5] The experimental sciences are given anything but priority in the latest primary school programmes. On the contrary, brief titles, lack of detail, etc., everything suggests that they are being overlooked. In a first version of the lycée reform plan presented in June 2008, the basic second year curriculum includes no lessons in physics or chemistry, nor in the life and earth sciences.

[6] This plan has been dropped.

Worldwide Tributes

The death of Pierre-Gilles de Gennes aroused much emotion and there were many who wished to honour his memory. In the months after his death, numerous tributes were paid at seminars and conferences. Dozens of streets and lycées now bear his name. However, so great was his popularity that there were strong incentives to capitalise on his name. In fact, the boundary between authentic tribute and exploitation is sometimes blurred. When Valérie Pécresse cited Pierre-Gilles de Gennes as a model in her opening speech for the "Young Researchers" programme (intended to make research careers more attractive) in July 2007 — *"I think of Pierre-Gilles de Gennes, who would give credit to his young collaborators by cosigning articles"* — she was paying tribute, but her aim was also to justify her plan. In October 2007, *Rhodia* set up a Pierre-Gilles de Gennes Innovation Centre on their Pessac site, then a Rhodia Pierre-Gilles de Gennes prize for science and industry, with a total value of €200,000.[7] No one can question the sincerity of the gesture (remember that Pierre-Gilles de Gennes worked for more than 20 years with *Rhône-Poulenc*, and then with *Rhodia*),[8] but this does not alter the fact that the company image had much to gain from these initiatives.

There is now a foundation that bears his name, the "scientific cooperation foundation", created by the research programme law of April 18, 2006. As soon as the law was introduced, the Institut Curie, ESPCI and the École Normale Supérieure applied to form a RTRA (advanced research network), a label that marks a centre of excellence in research and entitles them to a grant of €20 million and to the legal status of a "scientific cooperation foundation" to manage that grant.

---

[7] This prize was awarded on May 15, 2008 by Valérie Pécresse to the British physicist Richard Henry Friend, from Cambridge University. *"I know, Professor, that your peers could not pay you a greater compliment than linking your name and your work with the seminal figure of Pierre-Gilles de Gennes"*, she congratulated him.

[8] See Chapter 10.

Pierre-Gilles had supported the plan: *"We need to give it full backing"*, he insisted to doubters. The *Foundation for Transdisciplinary Life Research* came into existence in March 2007. It is a seed fund for innovation (designed to finance innovative research projects), which has to seek private funding, otherwise it will be dissolved once the initial sum has been distributed to various research projects.

When the death of Pierre-Gilles de Gennes was announced, emotions ran high amongst the foundation's board members, most of whom knew him well. Hervé Le Lous, Chairman of the foundation, heir to the Fournier laboratory and CEO of *Juva Santé* and *Urgo*, also realised the match between the foundation's objectives and the messages conveyed by the French Nobel laureate. He therefore proposed renaming the foundation after him, though it might seem paradoxical to give the name of a physicist to a foundation created *"to finance research projects that will lead to concrete breakthroughs in the field of health"*. The structure became the *Pierre-Gilles de Gennes Foundation for Research* in August 2007. Pierre-Gilles de Gennes became its brand image, with a photograph and quotation by the Nobel prizewinner on the website homepage and corporate brochure, the establishment of a "Pierre-Gilles de Gennes" interdisciplinary PhD programme of international seminars, etc.

The foundation is undoubtedly in tune with the views of the physicist, who had always argued for the harnessing of the energies of Montagne Sainte-Geneviève [Translator's note: an area of Paris that attracts many of the capital's students and academics], the spirit of enterprise and for interchange between public and private research, but *"his name is a brand that is an immensely effective fundraising tool"*, acknowledges its chief executive. *"The foundation will not survive if it is not called the Pierre-Gilles de Gennes Foundation for Research."*

His Children

As he said following the announcement of the Nobel Prize, honours were less important to him than his children. He was very proud of all of them. He lauded the talents of Christian, an excellent doctor, and his responsiveness to his patients. He admired his daughter Dominique, a French-Russian translator, then teacher in a lycée, *"a difficult profession"*. He boasted the merits of his sculptor daughter Marie-Christine, with whom he liked to discuss and compare the creative process involved in research and in art.

The other children are younger. The eldest, Claire, is a researcher in a neurology laboratory at the University of California, Berkeley, having passed through the École Normale Supérieure 45 years after her father. It was no longer the place of exploration and freedom that he knew. *"These days, students are encouraged to begin a thesis before leaving school in order to increase their chances of getting a position at the CNRS"*, he mourned. Matthieu is a researcher in physics at Harvard University in the USA, having chosen to study at the École Polytechnique. Pierre-Gilles de Gennes attended his thesis presentation in 2005, on *"the fluctuations of stock market prices and the microscopic properties of amorphous solids"*, in which he referred to an article published in 1968 by his father. It was the first time he had heard his son speak, and he was proud of his performance. *"He was concentrated, serious, not pompous"*, he rejoiced. The third child in the family, Olivier, became interested in Asia, and went to art school. Finally, Marc, *"the baby of the family"*, as his father would always call him, went into *classe préparatoire* in 2008.

Pierre-Gilles regretted not having been as close to his grandchildren as he would have liked. He had fond memories of the expedition down the Nile in March 1995, organised by Annie, with their seven grandchildren, aged 8 to 13. He had been amused by the undisguised irritation of their fellow passengers on seeing this *"hord of excited children"* come aboard. He would also tell the story of one of his most frightening moments, in Croatia, in summer 1982. He had rented a big holiday villa for the whole family on a small island where his friend Chichko had a house. One afternoon, he had taken five of his grandchildren on a motorboat excursion. They had left the port, then the bay, when the engine broke down. He was unable to restart it and the craft began to drift out to sea. He had grabbed the oars, rowing so hard that his arms began to hurt. They were a long way from harbour. Fortunately, and to his great relief, he managed to make it back. A photo of his grandchildren always hung in his office.

Looking Forward

At receptions, how often did Pierre-Gilles de Gennes turn away from the VIPs to talk to young people? He gave them a great deal of attention, regretting that our western societies offer them only *"mediocre goals of comfort or leisure"* and calling for the revival of ideals such as the ethics of knowledge and global solidarity. *"It is clear that if we do not succeed in making solidarity something to which our younger generations can aspire, in thirty years we will find ourselves in a worldwide conflict, and we will have lost everything. But will [these ideals] be enough to move us towards a stable civilisation in the 21st century? The answer will be in the hands of our children"*, he argued.[9] One of his suggestions to young people was the crazy dream of eliminating the Sahara by creating a worldwide fund to combat desertification and desalinate water.[10] One dream or another, it was of no consequence, provided that the passion was there. *"The important thing is to put heart and soul into what you do"*, he asserted.

---

[9] *Les objets fragiles*, Plon, 1994.
[10] *L'Heure de vérité*, December 27, 1992.

Appendices

Family History

De Gennes is the name of a 10th century Anjou family that held Château de Gennes, a castle on the banks of the Loire between Angers and Saumur. *"They were minor gentry: my ancestors were members of parliament or local tribunals in Poitou"*, explained Pierre-Gilles de Gennes. There was nothing connecting him with this small town, apart from the grave of his uncle, the physician Lucien de Gennes, who had curiously chosen to be buried there. *"In the 19th century, my ancestors tried unsuccessfully to prove that we were descended from an Olivier de Gennes referred to in the Song of Roland. That Olivier probably came from Geneva and had nothing to do with our family"*, he admitted. The genealogical record from the 15th century onwards was fairly accurate, indicating different branches of the family in Poitou, in Brittany and in Anjou. In the Renaissance, poems had been written dedicated to de Gennes maidens. In the reign of Louis XIV, certain de Gennes had been admitted to court or obtained high positions. The most famous was Admiral Jean-Baptiste de Gennes, whose expedition is described in the book *Account of a voyage made in 1695, 1696 and 1697 along the coasts of Africa, the Straits of Magellan, Brazil, Cayenne and the Antilles Isles, by a squadron of the King's vessels, commanded by M. de Gennes*, written by a member of the crew, François Forger. At that time, the French Royal Navy was battling the English and Dutch fleets for control of the seas, in order to maintain colonial momentum. The Straits of Magellan were a strategic route for gold and silver from the mines of Peru. Jean-Baptiste de Gennes initiated a *Plan for a French colony at the Straits of Magellan*. He left La Rochelle with a flotilla of six vessels on 3 June 1695 and first headed for Africa, where he seized Fort James, an English possession in Gambia, before crossing the Atlantic and following the coasts of modern-day Brazil and Argentina. In February 1696, he reached the Straits of Magellan, where there is still a "de Gennes River" running into French Bay. However, weather conditions and lack of provisions forced him to turn back. The squadron sailed backup the same coast, and Jean-Baptiste de Gennes conquered what is now French Guyana. He was rewarded with the title of Count by a royal decree of 12 July 1698, and became Governor of Guyana. He settled by the Oyac River in Guyana but his slaves revolted in 1700 and sacked the estate. Jean-Baptiste de Gennes was also Governor-General of St Kitts in the Caribbean, which he lost to the English in 1704. Accused of high treason and cowardice, he was returning to France to face execution, but fled to England, where he died soon after. In the end, his name was reinstated by the King.

"I don't remember dreaming of the nobility as a child. I was just interested in this admiral, Jean-Baptiste, whose story I had read when very young", reported Pierre-Gilles de Gennes. He did not give a fig for being a count. "I did a little calculation. Serious research has shown that ten per cent of sons of married parents are not their supposed father's child. So after ten generations, you only have a thirty per cent chance of being descended from who you think you are. That rather casts doubt on these questions of genealogy... In any case, I find the aristocratic tradition childish", he mocked.

"In the French Revolution, my ancestors were obviously on the side of the chouans [Translators note: leaders of a pro-royalist revolt]. One of them took part in the landing of the royalist emigres arranged with the English on the Quiberon Peninsular in 1795. However, they were driven back and massacred by General Hoche. After this, the family found itself leaderless and dispossessed of its property. The descendants eked out a difficult existence for most of the 19th century. Then there was a revival, and my grandfather became an important physician." His grandfather lived in the style of an important doctor of the time and had two sons, Robert, the elder and father of Pierre-Gilles, and Lucien, his uncle, who both became doctors after somewhat different career paths. In 1911, Robert began his military service, which at the time

lasted three years. Because of the advent of the First World War, he actually spent seven years in the army as a military doctor. By contrast, Lucien de Gennes studied under better conditions. "My uncle had the reputation of being an incredible diagnostician. He became a great endocrinologist", reported Pierre-Gilles de Gennes.

The de Gennes were originally Catholic, but the family of Pierre-Gilles' mother, whose maiden name was Yvonne Morin-Pons, were Protestant. A passionate genealogist, she traced her family back to the 17th century in the Drôme, then in Switzerland. They were barbers before they were bankers, founding the important Veuve Morin-Pons Bank in Lyon. Amongst her ancestors was Jacques Bernard (1795–1890), son of a Protestant family of silkworm breeders in the Cévennes. He successfully went into silk trading, becoming a significant figure in Lyon and mayor of the town of Brotteaux, where he drained the marshes during the reign of Louis Philippe. A great lover of painting, he built up a collection which now hangs in Lyon Museum.

Find out more

(Chapter 4, pages 43 to 69, A4-1)
Adiabatic processes
In thermodynamics, the term "adiabatic" means: "no transfer of heat to the outside environment during a system change." In statistical mechanics, it is shown that, at the microscopic scale, this process is found in transformations

where the levels of quantified energy move very slowly, at low frequencies compared with the jump frequencies between levels, whilst the probability that the level is occupied remains constant.

(Chapter 4, pages 43 to 69, A4-2)
Electrochemical etching of germanium

Pierre Aigrain's idea was to make a germanium film by electrolysing (a chemical reaction that takes place when an electrical current is passed through a solution) a sample of the material. As the surface decomposes (on contact with the ions in the solvent), the sample of germanium becomes thinner. While it is still thick enough, the current continues to get through, but once the thickness reaches a certain value, equivalent to the Schottky barrier, i.e. the electron-free area near the surface of the semiconductor, the current can no longer flow and the reaction stops: the sample has become an insulating film. Pierre-Gilles de Gennes had to conduct the experiment in the dark, because light creates current carriers, which would prevent the Schottky barrier acting as an insulator.

(Chapter 5, pages 71 to 89, A5-1)
Magnetic moments and magnetism in solids

Broadly speaking, magnetism is produced by the motion of electrical charges. Thus, the magnetic moments of atoms result from the motion of their electrons (in the classical model), firstly their movement in the electron cloud (so-called orbital magnetism), and secondly their rotation (so-called spin magnetism), which corresponds to the magnetic moment specific to electrons, called their "spin".

The magnetic moments of atoms are responsible for the magnetism of materials, which are classed into three main categories: diamagnetic (lead, silver, copper, etc.), paramagnetic (chromium, aluminium, platinum, etc.) and ferromagnetic (iron, cobalt, nickel, etc.). Put simply, when the atoms possess no permanent magnetic moment, the material is diamagnetic (diamagnetic materials have negative susceptibility, which arises from the tendency of the electrical charges to form a screen between the interior of the substance and the field: unlike the other two, diamagnetic materials are (very slightly) repelled by the field). When the atoms possess a permanent magnetic moment (due to the existence of unpaired electrons), the material is paramagnetic or ferromagnetic. If the moments are randomly orientated, the resulting magnetisation is zero and the material is paramagnetic (it becomes weakly magnetic when exposed to a field and does not retain its magnetism when the field disappears). In certain cases, however, interactions between the atoms prompt the magnetic moments to align themselves in parallel (within domains): the material is then ferromagnetic (it becomes magnetic through the effect of a magnetic field that aligns the different domains) and can retain its magnetism.

(Chapter 5, pages 71 to 89, A5-2)
Critical points
The liquid-gas transition is discontinuous and characterised by a density jump. However, as a critical point is reached (e.g. for water: 374.15°C and 218.3 atm), the shift from liquid to gas is continuous. The density fluctuations are gigantic. In the gas phase, there is a sort of fog of microdroplets which form, unform and scatter light: the fluid becomes milky ("critical opalescence").

(Chapter 5, pages 71 to 89, A5-3)
Spin wave spectrum of iron
Spin waves are the collective modes of a set of spins. These modes are characterised by the dispersion relation which links their frequency to the wavelength of the mode in question. Pierre-Gilles de Gennes tried to calculate the spin wave spectrum of iron, somewhat as the American physicist Richard Feynman had calculated that of "rotons", elementary excitations in helium 4.

(Chapter 5, pages 71 to 89, A5-4)
Isotope separation by running a
current through a molten metal
Maurice Goldman described the following situation to Pierre-Gilles de Gennes. When a strong current is run through a capillary filled with liquid gallium composed of two isotopes (two atoms that have the same number of protons and electrons, but not the same number of neutrons, so that they have different masses), after a few days the light isotope becomes concentrated near the anode. To explain this, Pierre-Gilles de Gennes imagined that

each gallium atom vibrated in a "cage" formed by its neighbours. Given that the vibrations depend on the atom's mass, it is as if *"the light isotope (was) slightly more "delocalised" than the heavy isotope (...): because its vibration amplitude is larger, it disrupts the motion of the electrons more, and experiences more collisions (and is carried along by them)"*, he wrote. That is why the light isotope is dragged along more by the electrons.

(Chapter 5, pages 71 to 89, A5-5)
Magnetism in rare earths
Of the fourteen rare earths (also called lanthanides), a dozen are ferromagnetic, some of them even more so than iron itself. The electronic structure of the rare earths differs from one element to another only in the number of electrons in the deep "4f" subshell. Pierre-Gilles de Gennes explained why these elements sometimes showed a high degree of ferromagnetism by the interactions between the f subshells of the atoms which bring conduction electrons into play (so-called Ruderman-Kittel interactions) and which are similar from one element to the next: only the magnitudes of the spin and of the orbital moment vary. He then established a formula which predicts the critical temperature of ferromagnetic transition on the basis of the atomic parameters of each element, in other words the orbital moment and the spin moment.

(Chapter 5, pages 71 to 89, A5-6)
Helicoidal magnetism
A magnetic system can be ferromagnetic if the spins are all aligned in the same

direction, or antiferromagnetic if the spins are aligned in antiparallel. A third, helicoidal order, where the spins rotate, appears if the interaction between first neighbours is ferromagnetic and between second neighbours is antiferromagnetic, as Jacques Villain showed.

(Chapter 5, pages 71 to 89, A5-7)
Suhl interactions
As nuclear spins and electronic spins are coupled, when a nuclear spin tilts, the nearby electronic spin also tilts, and as it is coupled to the other electronic spins, it eventually has a long-distance effect on the nuclear spins. These indirect interactions between nuclear spins bear the name of Harry Suhl.

(Chapter 6, pages 91 to 103, A6-1)
Double-exchange mechanism
An interesting mechanism occurs in weakly conducting magnetic materials where a few rare electrons migrate into the crystal and encourage the neighbouring moments to come into alignment. Broadly speaking, in this mechanism the electron of one ion, for example manganese Mn^{3+}, is transferred to the oxygen ion, which in turn passes it to another manganese ion Mn^{4+} (so there is a double exchange). This phenomenon, described by the American physicist Clarence Zener in 1951, explains why certain mixed-valence compounds (e.g. those containing both Mn^{3+} and Mn^{4+} manganese ions) are electrical conductors and ferromagnetic, despite the absence of unpaired electrons (usually responsible for conduction). Generalising this model in 1955, Philip Anderson had

highlighted a mismatch between theory and experimental data at high temperature, which Pierre-Gilles de Gennes resolved by showing that this difference only appeared in one particular case (when the width of the conduction band was greater than the thermal energy). He also refined the description of the double-exchange mechanism (conducive to "tilted" phases where the magnetic moments carried by the ions deviate from their axis) and defined the conditions when it occurs. The theorist would return to this mechanism in 1987, proposing an analogy between manganese oxides and mixed-valence superconducting copper oxides.

(Chapter 6, pages 91 to 103, A6-2)
Spin waves in garnets
Although weak, Suhl interactions cause long-distance effects and every nuclear spin interacts with millions of others. For this reason, and although the nuclear spins are disordered, collective modes (spin waves) arise, which would be experimentally demonstrated by University of California, Santa Barbara's Alan Heeger (awarded the 2000 Nobel Prize for chemistry for the discovery of conducting polymers).

(Chapter 7, pages 105 to 131, A7-1)
Fritz London and superconductivity
Fritz London, a German theorist exiled to England, showed that when a superconductor is immersed in a magnetic field, currents appear at its surface (at a depth called the London penetration depth), thereby creating a magnetic field in the sample which counterbalances

the external field. The result is that the field within the superconductor becomes zero. The German physicist likened the superconducting sample to "a single large diamagnetic atom", diamagnetism being a property of all materials in which the magnetic moment produced by an external magnetic field tends to oppose that field (A5-1). London interpreted superconductivity as a macroscopic quantum phase. By analogy with superfluidity in helium 4, his idea was that superconductivity resembles a Bose-Einstein condensate, in which the bosons are precipitated into a single (and macroscopic) state which experiences no dissipation mechanisms.

(Chapter 7, pages 105 to 131, A7-2)
Phase transitions
A phase transition is a phase change that occurs in a system when an external parameter is continuously varied, for example ice turning into water when the temperature is increased. Phase transitions are described as being of first or second order. First-order transitions are called discontinuous, because they show a discontinuity in the first derivative of the free energy in the system (presence of latent heat). This essentially means that they occur suddenly, for example at a set temperature in the case of melting ice. Second-order, so-called continuous phase transitions, on the other hand, are gradual. In ferromagnetic transition, for example, the magnetic moments gradually lose their shared orientation as temperature increases. Here, according to Landau's phase transition theory, a parameter

called the order parameter gradually changes as the transition approaches and vanishes at the transition point, following a temperature-based power law (in which the exponent is called the critical exponent). So in ferromagnetic transition, the order parameter is magnetisation: as the temperature increases, the spins gradually lose their orientation, with the result that the magnetisation progressively decreases, finally falling to zero at the Curie point.

(Chapter 7, pages 105 to 131, A7-3)
Shubnikov phase and vortices
The Russian physicist Alexei Abrikosov used the Ginzburg-Landau equations to show that there are two types of superconductors. Type I experience a sudden transition when the magnetic field reaches a critical value, whereas type II gradually lose their superconductivity between a field value HC1 and another called HC2. Between HC1 and HC2, in the so-called Shubnikov phase, the superconductor gradually allows the magnetic field to penetrate via tubes formed by vortices of superconducting currents which prevent the magnetic field reaching the rest of the superconductor. These tube-like vortices form a regular lattice.

(Chapter 7, pages 105 to 131, A7-4)
Cooper pairs
How can two electrons be bound together in a Cooper pair? It works as follows: when an electron passes through the lattice, it exercises a slight attraction on the positive ions, which, before returning to their initial position, create a positive zone which then

attracts another electron. As a result, the electrons become bound and form so-called Cooper pairs, which constitute an "electron flow" capable of travelling without loss of energy.

(Chapter 7, pages 105 to 131, A7-5)
Superconducting tunnel junctions
Pierre-Gilles de Gennes equipped his laboratory with evaporators to prepare thin superconducting layers and employed a measurement technique previously unknown in France, based on Ivar Giaever's discovery of quantum tunnelling between superconductors in 1959. With this technique, characteristics of the superconductor can be measured directly, for example its forbidden energy zone and the electron density of states. Like semiconductors, superconductors have a forbidden zone, i.e. a range of energies that the electrons cannot adopt (they are forbidden by the laws of quantum physics).

To make the measurement, a layer of metal is oxidised and a layer of superconductor placed on it. When a voltage is applied on either side of this sandwich of metal/insulating oxide/superconductor, the electrons can pass through the insulating oxide layer by quantum tunnelling and generate a tunnel current, which depends on the electron density of states. As the forbidden zone resembles a "hole" in this density of states, it is directly visible in the junction's current-voltage characteristic. This tunnel junction technique is therefore used to access superconductor parameters that are crucial for testing the theoretical models (varying

the composition and thickness of the films modifies characteristic lengths in the superconducting films, which makes it possible to test these models).

(Chapter 7, pages 105 to 131, A7-6)
Vortices and neutron scattering
Pierre-Gilles de Gennes picked up Abrikosov's research, which proposed that the vortices that appear in the Shubnikov phase form a regular lattice. He described their structure using the Bogoliubov-de Gennes equations, and he realised that the neutrons, being sensitive to the magnetic field, could be diffracted not by the magnetic moments as usual, but by the regular structure of the vortices in the Shubnikov phase. He and Jean Matricon calculated the effective cross-section of neutron scattering by the vortices, and predicted that it should be measurable.

(Chapter 7, pages 105 to 131, A7-7)
HC3 field
When subjected to a magnetic field, type II superconductors repel it until the HC1 point. After this, the field penetrates via the vortices, which become increasingly numerous as the field increases. When it reaches HC2, the vortices invade the entire volume of the superconductor, which reverts to its normal metallic state. Nonetheless, a thin layer of superconductor remains on the surface until the HC3 point. When the field exceeds HC3, it disappears.

(Chapter 7, pages 105 to 131, A7-8)
Turbulence and drag
"Every time water flows at high speed, for example around the arches of a

bridge, whirlpools appear, then break up to become smaller whirlpools, then smaller again, etc., until viscosity "kills" them at the smallest scales. If you add a polymer to the water to increase its viscosity, you increase the friction and — presto! — it reduces the drag. In our new model, elastic effects dominate the viscosity effects, but I don't know if it has been verified", explained Pierre-Gilles de Gennes.

(Chapter 7, pages 105 to 131, A7-9)
Mean-field (or molecular field) approximation

The interactions between objects in a system, for examples the spins in a ferromagnetic material or the molecules in a liquid crystal, etc., can be described by the action of a mean field made up of all the bodies acting on a single object in the system. The effects caused by the presence of the objects are in a sense averaged out.

(Chapter 8, pages 133 to 148, A8-1)
Carr-Helfrich mechanism

Molecules in the liquid crystal undergo thermal agitation which alters their direction. They then form kinds of waves between the two electrodes. When a sufficiently intense electrical field is applied, because liquid crystals are highly anisotropic (not all directions are equivalent), especially for electrical conductivity, currents arise which accumulate charges at the peaks of the waves. The field exercises a force on the charges which sets the fluid in motion, dragging the molecules with it, resulting in the emergence of cells (the Carr-Helfrich mechanism).

(Chapter 8, pages 133 to 148, A8-2)
Thermal instabilities in liquid crystals

In liquid crystals, upward convection deforms the liquid crystal by gently moving and tilting the molecules. Because heat propagates more easily along the molecules than across them (as does electrical current), it accumulates in certain areas (as do electrical charges), which in turn alters the convection movement.

(Chapter 8, pages 133 to 148, A8-3)
Analogy between smectics A and superconductors

The wave function described by smectics is similar to that described by Cooper pairs in a superconductor, its amplitude representing the modulation in the density of the liquid crystals and its phase the position of the layers. Pierre-Gilles de Gennes thus showed that the transition temperature diminished when the liquid crystal was under constraint, just as it decreases in a superconductor subjected to a magnetic field. This similarity between phase transitions in liquid crystals and superconductors was also discovered by William McMillan, a physicist at Urbana in the US, at the same time as Pierre-Gilles de Gennes. But William McMillan's approach was microscopic, whereas Pierre-Gilles de Gennes' was based on the study of singularities present over greater distances. Pierre-Gilles de Gennes predicted the existence of a new phase (the so-called TGB — Twist Grain Boundary — or "blue phase"), a twisted smectic phase, analogous to the Shubnikov phase in type II superconducting transitions: because of a sort of competition between the lamellar shape and the tendency to twist, the twist penetrates the lamellar structure through defects, in the same way as magnetic flow infiltrates into type II superconductors via vortices.

(Chapter 10, pages 159 to 185, A10-1)
Superfluidity of helium 3
In autumn 1973, Vinay Ambegaokar asked Pierre-Gilles de Gennes about surface effects on the superfluid properties of helium 3. Two years earlier, physicists at Cornell had shown that this helium isotope became superfluid at very low temperature, in other words it flowed with zero viscosity. The two physicists wrote an article on the subject, where they transposed the Ginzburg-Landau description and equations from liquid crystal mechanics to describe one of the phases of helium 3. This provides an analogy between liquid crystals and superfluid helium.

(Chapter 10, pages 159 to 185, A10-2)
Vesicle merger
When two cells come into contact, an opening forms in their membranes. It gradually widens until they merge into a single cell. Using simpler objects (vesicles), Pierre-Gilles de Gennes showed that the merger depends on the difference in osmotic pressure between the interior and exterior of the vesicle. When this pressure is high, the surface tension increases and encourages the creation of small openings. Chemists confirmed the laws predicted by Pierre-Gilles de Gennes on difference in osmotic pressure

(obtained by changing the salinity of the external environment).

(Chapter 11, pages 187 to 209, A11-1)
Renormalization group
The close analysis of scale changes and variations in physical magnitudes that result from the so-called renormalization group, published by Kenneth Wilson in 1972, can be used to understand the behaviour of materials (magnetic, binary blends, etc.) close to a second-order phase transition (A7-2). Certain magnitudes follow power laws in that they vary with the difference at the critical temperature. This analysis can be used to calculate the corresponding critical exponents and to establish what are called universality classes of the critical phenomena, which depend on the dimension of the space ($d = 3$ for our universe, $d = 2$ for a surface, etc.) and on the dimension of the order parameter ($d = 1$ for a scalar, $d = 3$ for a vector, etc.).

(Chapter 11, pages 187 to 209, A11-2)
n = 0
When Pierre-Gilles de Gennes applied the renormalization group to tangled polymers, he "cut up" the chain into small sections a few monomers long, characterised them, then grouped them into longer units. He repeated the operation until he had reconstituted the full chain. He went through a stage in which the dimension of the space was four, at which point the mean field approximation became correct. He then reduced the dimension of the space, and realised that if the order parameter was given a zero number of components ($n = 0$) (for example,

in a ferromagnetic-paramagnetic transition, the order parameter — magnetisation — has three components in space, but two components if it rotates in a plane), the phase transition diagrams became similar to the characteristic polymer diagrams. From this, he rigourously deduced the exponents of the power laws followed by polymers, drawing on the science of phase transitions for the tools needed to resolve the outstanding questions. Despite appearances, an order parameter with a component of zero has a physical meaning: it reflects the fact that the chain does not cross over itself.

(Chapter 11, pages 187 to 209, A11-3)
Semi-dilute polymer solutions

In a dilute solution, the polymer chains swell and behave as if they were isolated. When the concentration is increased, the chains tangle to form a sort of net in which the mesh size is the "correlation length". To model the properties of polymers in a semi-dilute solution, Pierre-Gilles de Gennes used the concept of blobs (where the size of the blob is the correlation length): within a blob, the chain is swollen and behaves like an isolated chain. By assuming that the blobs are compact, the correlation length can be deduced from the concentration. Then, by applying Flory's argument (which states that in molten solutions, the polymer chains return to the ideal state (A11-4)) to the blob scale, it emerges that the chains in a semi-dilute solution are ideal chains of blobs. The osmotic pressure of a semi-dilute solution is proportional to the density of the blobs.

(Chapter 11, pages 187 to 209, A11-4)
Molten polymer

The American physical chemist Paul Flory proposed that the chains in a molten polymer become ideal. The interactions are strong, but the potential is flat (because the concentration is uniform), with the result that the excluded volume is completely screened (the correlation length is then the size of a monomer). This was verified by neutron scattering.

(Chapter 11, pages 187 to 209, A11-5)
Plato and scaling laws

Were scaling laws known in antiquity? Gérard Jannink, a specialist in neutron scattering at CEA Saclay and a close colleague of Pierre-Gilles de Gennes, pointed out to him in a 2004 letter that a passage in *The Republic* shows that Plato had formulated a scaling law.

Socrates: Justice, which is the subject of our enquiry, is, as you know, sometimes spoken of as the virtue of an individual, and sometimes as the virtue of a State.

Adeimantus: True.

Socrates: And is not a State larger than an individual?

Adeimantus: It is.

Socrates: Then in the larger, the quantity of justice is likely to be larger and more easily discernible. I propose therefore that we enquire into the nature of justice and injustice, first as they appear in the State, and secondly in the individual, proceeding from the greater to the lesser and comparing them.

(Chapter 11, pages 187 to 209, A11-6)
Stabilisation of suspensions by
polymers
Colloidal particles (particles smaller than a micro-metre) stick together through Van der Waals forces, forming clusters that sediment (or "cream", depending on the density). These suspensions can be stabilised by preventing the particles getting close to each other and thus combating these Van der Waals forces. Grafting or absorbing polymer chains in the presence of a good solvent creates a "corona" of monomers which keeps the particles apart (steric repulsion). If these particles are saturated with polymers, they repel each other. If not enough polymers are introduced, a chain can form a bridge between two particles and the suspension becomes flocculated. Whether or not a colloidal solution is stabilised or destabilised, therefore, depends on the concentration.

(Chapter 11, pages 187 to 209, A11-7)
Plastic welding
When two thermoplastic blocks at their melting point are held together, the chains diffuse, sliding between each other by reptation, so that after a time the two blocks become welded. Pierre-Gilles de Gennes showed that polymer diffusion was very different from classical diffusion, because it led to step concentration profiles. The diffusion of identical chains was slow, but faster when the chains were different and showed an affinity for each other. This modified the profiles. Françoise Brochard-Wyart did the calculations and ended with a line segment profile, which seemed improbable for such

complex effects. Pierre-Gilles de Gennes was amazed, but it was in fact the strange outcome of the calculation of non-linear diffusion.

(Chapter 11, pages 187 to 209, A11-8)
"Coil-stretch" transition
Pierre-Gilles de Gennes discovered and characterised a phase transition in dilute polymer solutions subjected to an elongational flow, ideas developed during his stay in Bangalore, India, in November 1972. When the velocity gradient reaches a certain speed, the polymer coil suddenly uncoils into a long, stretched chain. This so-called "coil-stretch" transition is a first order phenomenon and predicts the existence of a hysteresis effect: the chain remains stretched even when the velocity gradient is reduced below the threshold value. For a long time, there were doubts about the existence of this phase transition.

One of the doubters was Bob Bird, from the University of Wisconsin, Madison, who had interminable discussions with Pierre-Gilles de Gennes at a conference in 1980. After 1985, experiments conducted notably at Bristol University demonstrated the coil-stretch transition and the associated hysteresis effect.

(Chapter 11, pages 187 to 209, A11-9)
Oil recovery
When injected, the water forms fingers that penetrate and slide through the oily mass, quickly connecting to the second well, with the result that it produces only water. The appearance of

these fingers is due to the Saffman-Taylor instability. If a low-viscosity fluid is pushed into a cell, the front is stable. Otherwise, the front is unstable and fingers form.

(Chapter 11, pages 187 to 209, A11-10)
Microemulsions
Pierre-Gilles de Gennes defined a characteristic length, called the persistence length, which represents the portion of the interface film that may be deemed flat (this persistence length depends on the curvature elasticity of the film). This allowed him to describe the interface as a random walk with this persistence as the step size. He then likened the microemulsion to a pile of small cubes, filled randomly with water or oil, with a ridge also the persistence length. Using this model, he explained under what conditions microemulsions are more stable than ordered phases. Broadly speaking, when the persistence length is small, the interface is very wrinkled, which encourages the appearance of sponge or bicontinuous phases (bicontinuous phases are like sponge phases in which the holes are also connected). When the persistence length is large, the interfacial films are flatter and tend to form stacks: the phase becomes lamellar. His model also predicted that the minimum quantity of surfactant needed to obtain a stable microemulsion is inversely proportional to the persistence length. Invented by Pierre-Gilles de Gennes, persistence length is now a fundamental parameter of microemulsion physics.

(Chapter 11, pages 187 to 209, A11-11)
Sol-to-gel transition
Pierre-Gilles de Gennes made use of the percolation model for the first time in the 1970s to describe the gelation of a monomer solution (sol-to-gel transition). The situation is as follows. Monomers are immersed in a solution. They are not bound, so the solution is fluid. Introducing an additive triggers a chemical reaction. The monomers bind together (polymerisation) and the solution becomes increasingly viscous as polymerisation develops. Once all the monomers are bound together, they form a giant molecule that extends across the container the solution has transformed into a gel. The physicist used the percolation model to determine the power laws that characterise sol-to-gel transition.

(Chapter 11, pages 187 to 209, A11-12)
Microemulsions with conducting droplets
Oil-rich microemulsions are insulators. But they can become conductors when enough water (and the right additives) are added for the conducting microdroplets of water to be numerous enough to join together. However, because of thermal agitation, varying sized conducting clusters quickly form and separate. For this reason, the classical percolation model does not work. The physicist Michel Laguës, who was then studying this conducting phase transition in microemulsions at the Collège de France, asked Pierre-Gilles de Gennes how the motion of the droplets could be accounted for, leading to the

latter's invention of the concept of "stirred percolation".

(Chapter 11, pages 187 to 209, A11-13)
Superconducting and percolation

Pierre-Gilles de Gennes also used the percolation model to study the presence of superconducting grains in a non-superconducting matrix (based on an idea by Guy Deutscher). These grains form clusters that have different magnetic properties depending on whether or not they form loops in which currents can flow. He extended his study to lasso-shaped superconducting wire (a loop with a branch) and showed that the conducting properties of these interwoven lassos depended on their connectivity rather than their shape. With his friend Shlomo Alexander, he constructed a theory of superconducting micro-networks, called the "de Gennes-Alexander" theory. In 2001, Pierre-Gilles de Gennes received an article accompanied by a note saying: *"You might be interested in this publication, which confirms your 20-year-old prediction of the existence of a destructive regime in small superconducting rings. This paper will be published in the next issue of* Science, *on 14 December."* He wrote back: *"It is a great pleasure to see old ideas become a reality."*

(Chapter 11, pages 187 to 209, A11-14)
Plug flow

Percolation can be used to describe flows of particles that are big enough for thermal agitation to be negligible. When there are small numbers of such particles, the flow obeys the classical (Poiseuille) laws and adopts a parabolic velocity profile (with velocity high in the centre and zero at the walls). However, when the particle concentration is higher, the velocity profile becomes flat in the centre of the tube (so-called "plug" flow). Pierre-Gilles de Gennes explained this by imagining that the particles collide with each other in the flow and form clusters as a result of van der Waals attractive forces, which disrupt the rheology of the system. Once the particles reach a certain concentration, an infinite cluster of particles appears in the flow. By rewriting the hydrodynamic equations, he developed the plug flow profile and showed that the average size of the clusters defined the edge of the flow.

(Chapter 11, pages 187 to 209, A11-15)
Wetting

The parameter S that governs the spread of a drop on a substrate is the difference between the surface energy of the dry solid and of the solid when covered by a film of fluid. If $S > 0$, the fluid spreads to reduce the surface energy (total wetting). If $S < 0$, the liquid remains in drops (partial wetting), characterised by a contact angle. The sign of S can be controlled by depositing a molecular film on the substrate (e.g. grafting a hydrophobic molecular mat to make glass non-wetting). Whilst partial wetting had been understood since Young (1805), the film thickness in the case of total wetting, "van der Waals pancakes", was not described until 1985 by Jean-François Joanny and Pierre-Gilles de Gennes. In the dynamics of total wetting, a precursor film develops around the drop. François

Heslot and Anne-Marie Cazabat showed that the drop spreads by forming stepped layers, like Aztec pyramids. The liquid behaves like a smectic liquid crystal, although it is an anisotropic phase.

(Chapter 11, pages 187 to 209, A11-16)
Ageing of a film of soap
A soap film extracted vertically from a container of soapy water consists of two walls of surfactants with a film of water trapped between. It tends to sag under its own weight, but a force caused by the surfactants pulls it upwards. However, this force only applies to the walls formed by the surfactants, whereas gravity acts on the whole film. Result: the water flows slowly within the soap walls, thinning the upper film (which appears black) by draining and thickening the lower film. Pierre-Gilles de Gennes applied the mechanisms involved to foams, multitudes of small bubbles connected by channels in which the liquid flows.

(Chapter 11, pages 187 to 209, A11-17)
Adhesion: the "peel test"
Pierre-Gilles de Gennes showed that the quality of a glue was characterised by the separation energy, which depends not only on the energy of the interfaces created at the fracture, nor on the energy of the fractured chemical bonds, but above all on the stretching of the polymers. When pulled, the glue joint deforms and fractures and the fracture then propagates within the joint. The constraint is highest at the top of this fracture where, as a result, the tangled polymers, like interwoven elastic bands, transmit the constraint to the other polymers. As they stretch, those closest to the top of the fracture disentangle, break in their turn and collapse: the fracture grows. As they stretch, the polymers dissipate energy. *"When you stretch a rubber band and suddenly let go, the band snaps back on your fingers. It has stored up energy, which it then restores, sometimes painfully. It is the same thing with polymers"*, he explained. That is why the softer elastomers are, the stickier they are.

(Chapter 12, pages 211 to 229, A12-1)
High-temperature superconductivity
High-temperature superconducting copper oxides have a similar structure to the mixed valence manganese oxides that Pierre-Gilles de Gennes had studied at Berkeley in 1960. These manganese oxides are unusual in that they are conductive despite the absence of free electrons, thanks to the so-called double-exchange effect (A6-1). The theorist wondered whether this effect might not explain superconductivity in copper oxides, by causing a gap in the (tilted phase) spin sub-lattices that would allow electrons to jump easily to neighbouring sites. They would then interact so strongly via the magnetic lattice that this attraction would not only be likely to generate Cooper pairs, but would also be sufficiently high for the critical superconductor temperature to be around the Néel temperature, i.e. the temperature at which magnetic order is lost. He wrote an article along these lines in the *Proceedings of the Academy of Sciences*, which he sent to the American physicist

Vinay Ambegaokar. Unknown to him, his article caused some excitement at Cornell University. Had Pierre-Gilles de Gennes found the key to the mystery? The American scientists were unable to satisfy their curiosity immediately, because the article was... in French! Vinay Ambegaokar asked one of his Canadian students, André-Marie Tremblay, to translate it. The latter realised that the suggested model, linked with the double-exchange mechanism, assumed that the oxides were ferromagnetic — i.e. with spins all pointing in the same direction — before any chemical change. However, the superconducting oxides were initially antiferromagnetic (the direction of the electron spins alternated from one atomic site to another) and became superconducting when their chemical composition was altered (by doping) in such a way as to allow an average of slightly less than one electron per atomic site to move. Pierre-Gilles de Gennes then proposed taking into account the crystalline-field effect, i.e. the idea that the environment of the copper atoms restores the antiferromagnetic state, and published an article describing this model. In the end, he rejected the idea that the double-exchange mechanism could be involved, and that oxide superconductivity could be a form of contrary antiferromagnetism, the most widespread hypothesis at the time (because copper oxides were antiferromagnetic, many physicists were looking to the arrangement of spins for the key to high-temperature superconductivity).

(Chapter 13, pages 231 to 262, A13-1)
Nobel Prize selection
Every year, the Nobel Committee for physics, which is part of the Swedish Royal Academy of Sciences, invites 3000 scientists around the world to make their nominations. The committee makes a shortlist of the 300 or so names suggested each year and asks physicists for comprehensive reports on the researcher's career. This does not guarantee that the person will win the prize, but proves that he or she is definitely in the running. The prize is attributed by the committee in October, by a majority vote.

(Chapter 13, pages 231 to 262, A13-2)
Nuclear fusion
Nuclear fusion occurs when atomic nuclei reach temperatures in excess of 10 million degrees and combine: the reaction releases considerable energy which, if recovered, could be used to produce electricity. Since no material can withstand such temperatures, the nuclei are held suspended in a magnetic field created by superconducting magnets. Pierre-Gilles de Gennes did not believe that nuclear fusion might one day replace ordinary nuclear fission.

(Chapter 14, pages 263 to 276, A14-1)
Pressure distribution in a silo
In a silo, the column of grain sags (slightly) under its own weight. The problem is doubly complicated, firstly because the elasticity of the granular material depends on pressure (and therefore on the quantity of grain poured), and secondly because the

weight of the grain is partly supported by the vertical walls, where friction (between grains and against the wall) plays a role that is essential but difficult to evaluate. Pierre-Gilles de Gennes defined a so-called anchor length which is the minimum length for friction to occur, such that when the movement of the grains is less than the anchor length, this friction can be ignored. He also studied thermal variations in the silo, which can cause the grain mass to expand or shrink, and showed that their effects are localised at the top of the column.

(Chapter 14, pages 263 to 276, A14-2) BCRE avalanche model

The BCRE (Bouchaud, Cates, Ravi Prakash and Edwards) model is based on the following observation: when sand is poured onto a pile, there are grains that roll (near the surface) and grains that do not move (under the surface). In other words, there is a boundary between the fluid state (mobile grains) and the solid state (immobile grains). The model also takes into account the exchanges between mobile grains (which can be trapped) and static grains (which can be set in motion).

(Chapter 14, pages 263 to 276, A14-3) "Bouchaud-de Gennes" model

Pierre-Gilles de Gennes introduced a new, so-called neutral angle, which is the angle of the sand pile at which as many grains are dislodged (and therefore set in motion) as are trapped (and therefore motionless), at the time of the avalanche, which therefore maintains a constant volume during the flow.

(Chapter 14, pages 263 to 276, A14-4) Segregation

Pierre-Gilles de Gennes explained certain segregation phenomena, in particular in silos. When the size difference between grains is small, his model predicted that continuous segregation occurs in the static phase during filling of the silo. When the size difference is greater, grain segregation needs to be taken into account in the mobile phase and the model predicts complete segregation (the large grains separate from the small and alternating bands of large and small grains form).

Bibliography

Books

Anatole Abragam, *De la physique avant toute chose*, Odile Jacob, 1987.

Charles Ailleret, *L 'aventure atomique française*, Grasset, 1968.

Jacques Badoz et Pierre-Gilles de Gennes, *Les objets fragiles*, Plon, 1994.

André Bendjebbar, Histoire secrète de la bombe atomique française, Le Cherche-midi, 2000.

Nicolas Chevassus-au-Louis, Savants sous l'occupation: Enquête sur la vie scientifique entre 1940 et 1944, Le Seuil, 2004.

Mohammed Daoud and Claudie Williams, *Introduction à la matière molle: la juste argile*, EDP Sciences, 1995.

Jacques Duran, *Passion chercheur*, Belin, 2005.

Jacques Friedel, *Graine de mandarin*, Odile Jacob, 1999.

Pierre-Gilles de Gennes, *Petit point*, Le Pommier, 2002.

Pierre-Gilles de Gennes, *Simple views on condensed matter*, World Scientific Publishing, 1992.

Bernard Jacrot, *Des neutrons pour la science*, EDP Sciences, 2006.

Promod Kumar *et al.*, Handbook of microemulsion science and technology, CRC Press, 1999.

Laurent-Patrick Lévy, *Magnétisme et supraconductiité*, EDP Sciences, 1997.

Jean Matricon and Georges Waysand, *La guerre du froid*, Le Seuil, 1994.

Michel Mitov, *Les cristaux liquides*, PUF, 2000.

Sven Ortoli and Jean Klein, *Histoire et légendes de la supraconduction*, Calmann-lévy, 1989.

Alain Peyrefitte, *Rue d'Ulm*, Fayard, 1994.

Richard J. Weiss, *A physicist remembers*, World Scientific, 2007.

Articles

Monique Brunet, « Cristaux liquides: histoire de la recherche en France », *Photoniques*, 2003.

Pierre-Gilles de Gennes, « Les colles font de la résistance », *La Recherche*, janvier 1997.

Pierre-Gilles de Gennes, « La percolation: un concept unificateur », *La Recherche*, novembre 1976.

Françoise Ploquin, « Une éducation de bon sens », *Le français dans le monde*, janvier 2005.

Marie Foucard, « Les découvertes géographiques et la perception de l'espace maritime à la fin du xviie siècle. L'exploration des terres inhabitées du détroit de Magellan (1 698-1 701) », *Hypothèses*, 2001.

Groupe des cristaux liquides d'Orsay, « Les cristaux liquides », *La Recherche*, mai 1971.

Chantal Houzelle, « Recherche: le cri d'alarme d'un prix Nobel », *Les Échos*, 12/01/2006.

Marie-Jeanne Husset, « Nous ne sommes pas des oracles », *Sciences et avenir*, décembre 1991.

Liliane Léger, « Pierre-Gilles de Gennes et l'aventure de la matière molle », *Images de la physique*, CNRS éditions, 2007.

Giuseppe Marrucci, « Obituary for Pierre-Gilles de Gennes », *Rheologica Acta*, 47, 1, 2008.

Anita Mehta, « As a bracelet melts into gold... », *The Hindu*, May 31, 2007.

Hari Prasad Sharma, « Type-II superconductors, vortex lattice and role of an Indian scientist », *Current Science*, vol. 93, N° 9, 2007.

Michel Soutif, « La connivence entre physiciens de 1950 à 1975 », *La revue du CNRS*, mai 2000.

Un savant nommé Pierre-Gilles de Gennes, Science et Vie hors série, N°192, 1995.

Index